Theoretical and Observational Problems Related to Solar Eclipses

NATO ASI Series

Advanced Science Institutes Series

A Series presenting the results of activities sponsored by the NATO Science Committee, which aims at the dissemination of advanced scientific and technological knowledge, with a view to strengthening links between scientific communities.

The Series is published by an international board of publishers in conjunction with the NATO Scientific Affairs Division

A	Life Sciences	Plenum Publishing Corporation
B	Physics	London and New York
C	Mathematical and Physical Sciences	Kluwer Academic Publishers
D	Behavioural and Social Sciences	Dordrecht, Boston and London
E	Applied Sciences	
F	Computer and Systems Sciences	Springer-Verlag
G	Ecological Sciences	Berlin, Heidelberg, New York, London,
H	Cell Biology	Paris and Tokyo
I	Global Environmental Change	

PARTNERSHIP SUB-SERIES

1.	Disarmament Technologies	Kluwer Academic Publishers
2.	Environment	Springer-Verlag / Kluwer Academic Publishers
3.	High Technology	Kluwer Academic Publishers
4.	Science and Technology Policy	Kluwer Academic Publishers
5.	Computer Networking	Kluwer Academic Publishers

The Partnership Sub-Series incorporates activities undertaken in collaboration with NATO's Cooperation Partners, the countries of the CIS and Central and Eastern Europe, in Priority Areas of concern to those countries.

NATO-PCO-DATA BASE

The electronic index to the NATO ASI Series provides full bibliographical references (with keywords and/or abstracts) to more than 50000 contributions from international scientists published in all sections of the NATO ASI Series.
Access to the NATO-PCO-DATA BASE is possible in two ways:

– via online FILE 128 (NATO-PCO-DATA BASE) hosted by ESRIN,
Via Galileo Galilei, I-00044 Frascati, Italy.

– via CD-ROM "NATO-PCO-DATA BASE" with user-friendly retrieval software in English, French and German (© WTV GmbH and DATAWARE Technologies Inc. 1989).

The CD-ROM can be ordered through any member of the Board of Publishers or through NATO-PCO, Overijse, Belgium.

Series C: Mathematical and Physical Sciences – Vol. 494

Theoretical and Observational Problems Related to Solar Eclipses

edited by

Zadig Mouradian

Observatoire de Paris,
France

and

Magda Stavinschi

Astronomical Institute,
Bucharest, Romania

Springer Science+Business Media, B.V.

Proceedings of the NATO Advanced Research Workshop on
Theoretical and Observational Problems Related to Solar Eclipses
Bucharest, Romania
1–5 June 1996

A C.I.P. Catalogue record for this book is available from the Library of Congress

ISBN 978-0-7923-4619-7   ISBN 978-94-011-5492-5 (eBook)
DOI 10.1007/978-94-011-5492-5

Printed on acid-free paper

This workshop is dedicated to the memory of
Professor Călin POPOVICI
(1910-1977)
the founder of the Astrophysics Department of
Bucharest Observatory

TABLE OF CONTENTS

Session 3. : Low temperature structures in coronal environment.

Session4: Specific problems of solar eclipse observations.

Session 5. : Tasks for total solar eclipse of 11 August 1999.

Session 6.:Instrumental improvement for future observations.

Session 7.: Public education at eclipse and eye safety.

PREFACE

The NATO ARW on the problems of ground-based observations of Solar Eclipses was held in Sinaia (Romania) between 1 and 5 June 1996.

The Workshop was divided into seven sessions, in which 17 papers were given, by key speakers, along with 30 oral presentations. Additionally, 30 posters were presented. This issue contains only the invited and oral papers. The posters are to be published in a special issue of the Romanian Astronomical Journal.

The contributions were based on our present knowledge of solar corona physics and on the perspectives for future total eclipse observations, focussing especially on that of August 11, 1999, which will be the last eclipse of the century. The workshop sessions reviewed the results of past eclipse observations, coronal hot and cold structures, coronal heating, public education, and instrumental problems. At the end of the meeting a fruitful general discussion drew out problems to be studied and techniques to be used for forthcoming observations. Posters completed the workshop contributions.

The relation of the corona to the Sun was discussed, i.e. the energy and mass transfer between the chromosphere and corona. Concerning this mass transport we must mention the formation of prominences by coronal condensation in coronal cavities and the supply of mass to the corona by spicules. The other aspect of the Sun - corona relation is the unsolved question of the coronal heating mechanism, for which it is stressed that observations of fine structures are necessary. The basic question is: does heating occur in current sheets or in sheeths surrounding flux tubes?

After precise projects taking advantage of recent technological developments were outlined for the 1999 eclipse. The forecast of Solar Activity foresees the future maximum precisely in 1999, which enhances the importance of these eclipse observations.

The desire was expressed for close cooperation between groundbased eclipse observers and space instruments like SOHO and Yohkoh. Thus a JOSO (Joint Organisation for Solar Observations) working group was formed to coordinate the choice of targets to be observed by all the instruments used during the eclipse. The contacts established during the workshop will continue and broaden within JOSO. A number of Eastern European countries want to participate in the 1999 eclipse observations .

Problems of public education, in particular eye safety questions, were also raised during the workshop.

It must be mentioned that the reception and accommodation were excellent, and the co-sponsoring was highly effective.

Z. Mouradian and M. Stavinschi

Acknowledgments.

We would like to thank the sponsors of this workshop, namely the NATO Scientific Affaires Division, the Romanian Academy, the Romanian Ministry of Science and Technology, and the Paris Observatory.

We are grateful to the Scientific Committee, to Local Organizing Committee, and specially to Dr. G. Maris.

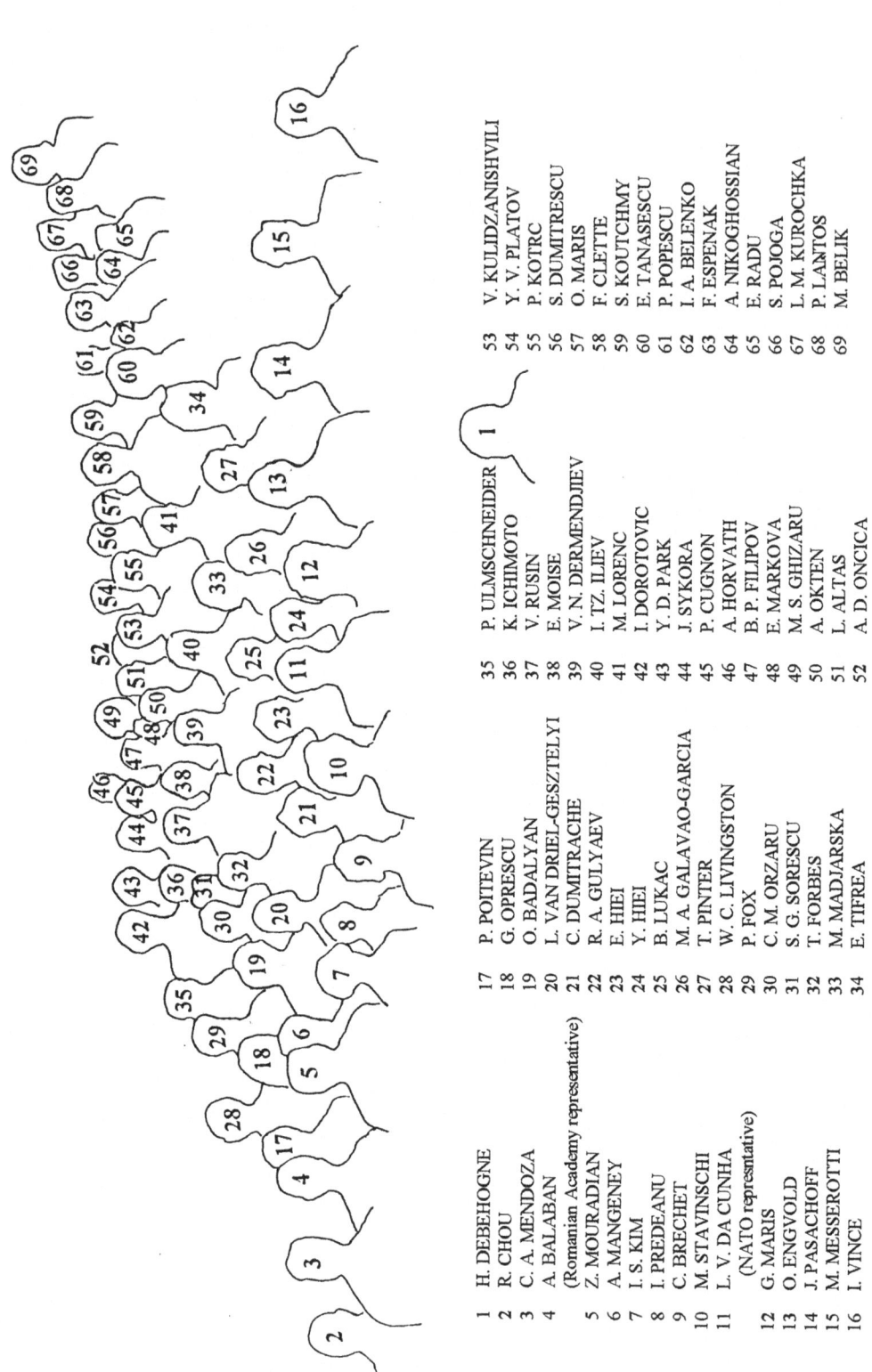

1	H. DEBEHOGNE	17	P. POITEVIN	35	P. ULMSCHNEIDER	53	V. KULIDZANISHVILI
2	R. CHOU	18	G. OPRESCU	36	K. ICHIMOTO	54	Y. V. PLATOV
3	C. A. MENDOZA	19	O. BADALYAN	37	V. RUSIN	55	P. KOTRC
4	A. BALABAN	20	L. VAN DRIEL-GESZTELYI	38	E. MOISE	56	S. DUMITRESCU
	(Romanian Academy representative)	21	C. DUMITRACHE	39	V. N. DERMENDJIEV	57	O. MARIS
5	Z. MOURADIAN	22	R. A. GULYAEV	40	I. TZ. ILIEV	58	F. CLETTE
6	A. MANGENEY	23	E. HIEI	41	M. LORENC	59	S. KOUTCHMY
7	I. S. KIM	24	Y. HIEI	42	I. DOROTOVIC	60	E. TANASESCU
8	I. PREDEANU	25	B. LUKAC	43	Y. D. PARK	61	P. POPESCU
9	C. BRECHET	26	M. A. GALAVAO-GARCIA	44	J. SYKORA	62	I. A. BELENKO
10	M. STAVINSCHI	27	T. PINTER	45	P. CUGNON	63	F. ESPENAK
11	L. V. DA CUNHA	28	W. C. LIVINGSTON	46	A. HORVATH	64	A. NIKOGHOSSIAN
	(NATO representative)	29	P. FOX	47	B. P. FILIPOV	65	E. RADU
12	G. MARIS	30	C. M. ORZARU	48	E. MARKOVA	66	S. POJOGA
13	O. ENGVOLD	31	S. G. SORESCU	49	M. S. GHIZARU	67	L. M. KUROCHKA
14	J. PASACHOFF	32	T. FORBES	50	A. OKTEN	68	P. LANTOS
15	M. MESSEROTTI	33	M. MADJARSKA	51	L. ALTAS	69	M. BELIK
16	I. VINCE	34	E. TIFREA	52	A. D. ONCICA		

WORKSHOP PARTICIPANTS

Directors:
Zadig MOURADIAN *Observatoire de Paris-Meudo FRANCE* mouradian@obspm.fr
Magda STAVINSCHI *Astronomical Institute, Bucharest ROMANIA* mstavinschi@roimar.imar.ro

Scientific Organizing Committee:
Oddbjorn. ENGVOLD *Inst. of Theoretical Astrophysic, Oslo NORWAY* oddbjorn.engvold@astro.uio.no
Georgeta MARIS *Astronomical Institute, Bucharest ROMANIA* gmaris@roimar.imar.ro
Jay PASACHOFF *Williams Coll. Hopkins Observatory, Williamstown USA* jmp@williams.edu

Local Organizing Committee:
Georgeta MARIS *Astronomical Institute, Bucharest ROMANIA* gmaris@roimar.imar.ro
Claudine BRECHET *Observatoire de Paris-Meudo FRANCE*
Alexandru DUMITRESCU *Astronomical Institute, Bucharest ROMANIA*
Mihai S. GHIZARU *Astronomical Institute, Bucharest ROMANIA* mghizaru@roimar.imar.ro
Sorin POJOGA *Astronomical Institute, Bucharest ROMANIA* spojoga@roimar.imar.ro
Gheorghe VASS *Astronomical Institute, Bucharest ROMANIA*

Participants:
Levent ALTAS *Kandilli Observatory, Istambul TURKEY* altas@boun.edu.tr
Olga BADALYAN *Izmiran Moscow Region RUSSIA* badalyan@izmiran.rssi.ru
Irina A. BELENKO *Sternberg State Astron. Inst.,Moscow RUSSIA* belenko@sai.msu.su
Marcel BELIK *Observatory Upice CZECH REPUBLIC* obsupice@mbox.vol.cz
Gheorghe-Dorin Gh. CHIS *Astron. Obs., Cluj-Napoca ROMANIA* 40.064.194.592 (Phon)
Ralph CHOU *School of Optometry,University of Waterloo CANADA* bchou@sciborg.uwaterloo.ca
Frederic CLETTE *Observatoire Royal de Belgique, Bruxelles BELGIUM* fred@oma.be
Pierre CUGNON *Observatoire Royal de Belgique, Bruxelles BELGIUM* pierre.C@oma.be
Henri DEBEHOGNE *Observatoire Royal de Belgique, Bruxelles BELGIUM* HenriDebohogne@oma.be
Vladimir N. DERMENDJIEV *Inst. of Astronomy , Sofia BULGARIA* vlderm @bgearn.acad.bg
Cristiana DUMITRACHE *Astronomical Institute, Bucharest ROMANIA* crisd@roimar.imar.ro
Ivan DOROTOVIC *Observatory Hurbanovo SLOVAKIA* rybansky@ta3.sk
Fred ESPENAK *NASA-GSFC, Greenbelt USA* espenak@gsfc.nasa.gov
Boris P. FILIPOV *Izmiran, Moscow Region RUSSIA* bfilip@izmiran.rssi.su
Thierry FORBES *Space Science Center, Durham USA* terry.forbes@unh.edu
Peter FOX *HAO - NCAR, Boulder USA* pfox@uluru.hao.ucar.edu
Jean-Rene GABRYL *Observatoire Royal de Belgique, Bruxelles BELGIUM* jean-rene@oma.be
Victor GAIZAUSKAS *Herzberg Inst. Astrophysics, Ottaw, CANADA* vgaizauskas@solar.stanford.edu
Maria Adriana GALAVAO-GARCIA *Obs. Astron., Coimbra PORTUGAL* astronomico@gemini.ci.uc.pt
Mihai S. GHIZARU *Astronomical Institute, Bucharest ROMANIA* mghizaru@roimar.imar.ro
Rudolf A. GULYAEV *Izmiran, Moscow Region RUSSIA* rgulyaev@izmiran.rssi.ru
Eijiro HIEI *Meisei University, Tokyo JAPAN* hiei@spot.mtk.nao.ac.
Kiyoshi ICHIMOTO *National Astronomical Observatory, Mitaka JAPAN* ichimoto@spot.mtk.nao.ac.jp
Ilko TZ. ILIEV *Solar-Terrestrial Influence Lab., Sofia BULGARIA* madjarsk@begearn.acad.bg
Iraida S. KIM *Sternberg State Astron. Inst.,Moscow, RUSSIA* kim@sai.msu.su
Pavel KOTRC *Astronomical Institute, Ondrejov CZECH REPUBLIC* pkotrc@asu.cas.cz
Serge KOUTCHMY *Institut d'Astrophysique de Paris FRANCE* koutchmy@iap.fr
Vazha I. KULIDZANISHVILI *Abastumani Observatory GEORGIA* kvazha@abao.kheta.georgia.su
Lev. M. KUROCHKA *Astron. Obs., Kyiv UKRAINA* lkur@aoku.gluk.apo.org
Pierre LANTOS *Observatoire de Paris-Meudon FRANCE* pierre.lantos@obspm.fr
William C. LIVINGSTON *NOAO-NSO,Tucson USA* wcl@noao.edu
Marian LORENC *Slovak Central Observatory, Hurbanovo SLOVAKIA* rybansky@ta3.sk
Bohuslav LUKAC *Slovak Central Observatory, Hurbanovo SLOVAKIA* rybansky@ta3.sk
Maria MADJARSKA *Inst. of Astronomy, Sofia BULGARIA* madjarsk@bgearn.acad.bg
Andre MANGENEY *Observatoire de Paris-Meudon FRANCE* mangeney@megasg.obspm.fr
Ovidiu MARIS *Gravitational and Space Sci. Inst., Bucharest ROMANIA* maris@ifa.ro
Eva MARKOVA *Observatory Upice CZECH REPUBLIC* markovae@mbox.vol.cz

Cesar A. MENDOZA *Centro Astrofisica Teorica, Merida VENEZUELA* cesar@ciensulave
Mauro MESSEROTTI *Astronomical Observatory, Triest ITALY* messerotti@ts.astro.it
Arthur NIKOGHOSSIAN *Byurakan Astrophysical Obs. ARMENIA* narthur@helios.sci.am
Adnan OKTEN *University Observatory,Istanbul TURKEY* istanbul@471tau.ege.edu.tr
Adrian D. ONCICA *Astronomical Institute, Bucharest ROMANIA* oncica@roimar.imar.ro
Carmen-Monica ORZARU *Astronomical Institute, Bucharest ROMANIA* morzaru@roimar.imar.ro
Young Deuk PARK *Astronomy Observatory KOREA* ydpark@flare.boao.re.kr
Teodor PINTER *Central Observatory SLOVAKIA* through:rybansky@ta3.sk
Yulii V. PLATOV *Izmiran, Moscow Region RUSSIA* bfilip@izmiran.rssi.su
Patrick POITEVIN *Flemish Astronomical Association BELGIUM* ppoitevin@innet.be
Sorin POJOGA *Astronomical Institute, Bucharest ROMANIA* spojoga@roimar.imar.ro
Irina PREDEANU *Astronomical Institute, Bucharest ROMANIA* ipredeanu@roimar.imar.ro
Eugenia RADU *Astron. Obs., Cluj-Napoca ROMANIA* 40 64 194 592 (Phon)
Voyto RUSIN *Astronomical Institute, Tatranska. Lomnica SLOVAKIA* vrusin@ta3.sk
Julius SYKORA *Astronomical Institute, Tatranska Lomnica SLOVAKIA* sykora@ta3.sk
Emilia TIFREA *Astronomical Institute, Bucharest ROMANIA* 40 1 337 33 89 (Fax)
Peter ULMSCHNEIDER *Inst.Theo. Astrophys., Heidelberg GERMANY* ulm@artemis.ita.uni-heidelberg.de
Lidia VAN DRIEL-GESZTELYI *Konkoly Observatory HUNGARY* lvandrie@mesioq.obspm.fr
Istvan I. VINCE *Astronomical Observatory Belgrade YUGOSLAVIA* ivince@aob.aob.bg.ac.yu

Results from the Coronal Observations of the 1994 and 1995 Total Solar Eclipse

E.Hiei, K.Inoue, N.Takahashi, and
Meisei University Eclipse Team
2-1-1, Hodokubo , Hino, Tokyo 191, Japan

Abstract. In order to know the three dimensional structure of the solar corona, collaboration of both the ground-based and soft X-ray observations is needed. Comparison of both observations at the 1994 and 1995 total solar eclipse is described, and their coronal streamers are discussed.

Key words. total solar eclipse, solar corona, coronal streamer,
soft X-ray observation, Yohkoh satellite.

1. Introduction

Collaborative observations at the 1994 and 1995 total solar eclipse were successfully made from the ground-based telescopes and the instruments on board the Yohkoh satellite. On 3 November 1994, the eclipse observations from the ground were made in Peru, Chili, Bolivia, Paraguay, and Brazil. It is reported that there was thin cirrus at Putre in Chili, but was fine in Paraguay and Brazil. The results of the 1994 eclipse were reported in Bolivia meeting (1995). Soft X-ray Telescope on board the Yohkoh Satellite (The Yohkoh Mission, 1991) has observed with a special observing programme for the 1994 total solar eclipse.

At the eclipse of 24 October 1995, the totality passed through Iran, India, Burma, Thai, Vietnam, and Kalimantan. The weather was good in India, but at Thai and Vietnam, the corona was observed through thin cloud, occasionally passed in front of the observers. The observations made in India will be published in Kodaikanal report

1

Z. Mouradian and M. Stavinschi (eds.),
Theoretical and Observational Problems Related to Solar Eclipses, 1-6.
©1997 *Kluwer Academic Publishers.*

(1996). The SXT (Soft X-ray Telescope) has also observed the 1995 eclipse event. Due to a rapid rotation period of 90 min of the Yohkoh satellite, the eclipse event was observed 4 times; 2 times for the coronal region out side of the eclipse, and 2 times at the eclipse.

Our observations, made at the 1994 eclipse in Paraguay (Hiei et al., 1995), and at the 1995 eclipse in India (Hiei et al., 1996), were compared with the SXT images, and foot region of coronal streamer is identified above a magnetic neutral line. The foot points of the polar plumes, however, are rather difficult to be identified on the SXT images.

2. Comparison of Soft X-ray corona and white light corona

The solar corona at a total solar eclipse can only be seen from its side view and the top view of the corona can be supplemented by the soft X-ray corona .

Fig.1 is the coronal image of the 1994 eclipse, taken at Vapor Cue in Paraguay. The image was processed by a computer in such a way that each coronal image is enhanced by a rotating unsharp masking method (Shiota, 1994), and then 5 different exposed images are overlapped. There appear a bright helmet streamer at the south-west limb, a wide streamer at the east limb, and a weak streamer at the north-west limb.

Figure 1. Coronal image of the 1994 eclipse, processed by a computer.

Synoptic maps, prepared by Mouradian et al. (1995) and by McIntosh (1995), show that magnetic neutral line, existing in the east-west direction, runs across the south-west limb, and an arcade above this neutral line is seen as the bright streamer. The data observed with the K-coronameter at Mauna Loa in Hawaii are helpful to the identification. At the east limb, a neutral line in the north-south direction is identified to be the foot region of the streamer at the east limb. The arcade structure along the neutral line is seen from its side, and thus the streamer at the east limb becomes wide. A coronal hole exists near the north-west limb, and the coronal magnetic field extends almost radially and the streamer at the northe-west limb is seen radially.

Fig.2 shows the corona and magnetic polarity region (McIntosh, 1995). Notice the neutral line crossing the south-west limb. Coronal holes are shown in dark regions. Coronal holes at the north- and south- polar regions are not symmetric to the solar axis, while the polar plumes appear symmetric, irrespective of the coronal holes. Numbers of the polar plumes near the base are shown in Table 1; there is a tendency that the numbers are a little larger in 1995 than in 1994.

Figure 2 . Magnetic neutral lines and coronal images.

TABLE 1. Number of polar plumes.

polar region	1994	1995
north	9	12
south	11	12

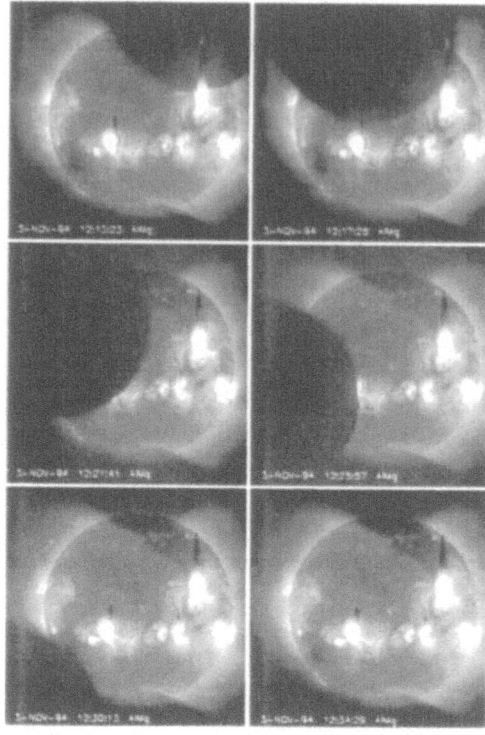

Figure 3 . Soft X-ray images at
the 1994 eclipse.

Fig.3 shows the soft X-ray corona taken with the SXT. The SXT image has been observing the corona of 3×10^6K plasma in wavelength of 2 - 70 Å (The Yohkoh Mission, 1991). The identification of the foot points of the polar plumes to soft X-ray bright points is difficult. It is to be noted that Fe IX / X images taken with the SOHO instrument show polar-plume like features and their bright foot points. Emission measure of the polar plumes is probably higher in 1×10^6K plasma in SOHO observations than 3×10^6K plasma in Yohkoh observations. Fe IX / X images taken simultaneously at the eclipse observations are premising to identify polar plumes and their foot points on the solar disk.

The observations on 24 October 1995 eclipse was made at Dundlodh in western part of India. Fig.4 shows the coronal image, processed by a computer as mentioned above. Synoptic map on the eclipse day shows that there are no neutral magnetic lines across the east limb. The streamer at the east limb is typical at a minimum phase of a solar activity. The streamer at the west limb seems to be weakly separated into two components. In fact the K-corona data at Mauna Loa shows two streamers at the west limb. There is a coronal hole at the west limb and thus the streamer appears rather radial.

Fig.5 shows a SXT image at the eclipse. The streamer at the west limb does not correspond to the active region near the west limb. X-ray bright points are seen, but it is not certain whether it corresponds to one of the foot point of the polar plumes.

Figure 4 . Coronal image of the 1995 eclipse, processed by a computer.

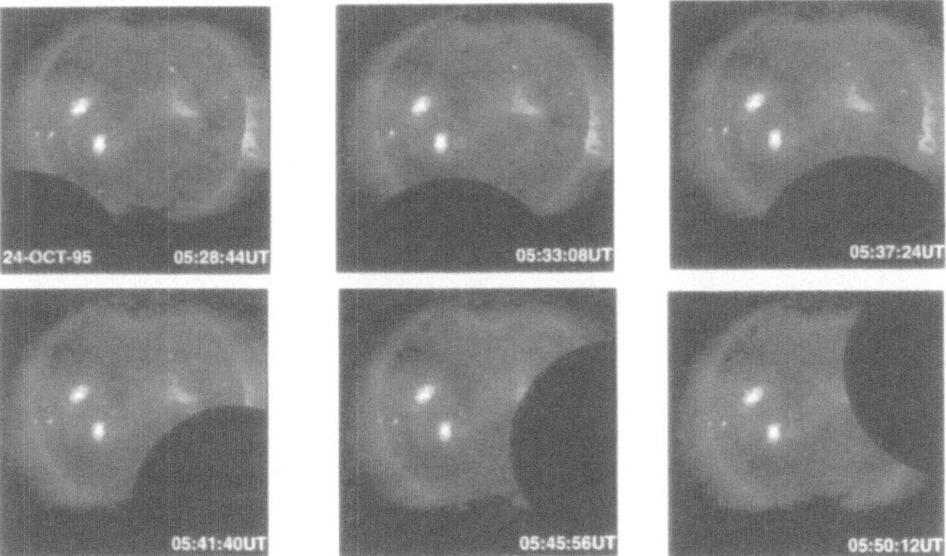

Figure 5. Soft X-ray images at the 1995 eclipse.

3 . Acknowledgements

This work was supported by Meisei Gakuen. We wish to thank K.Shiota for helping to make the 1994 eclipse image, and S.Watanabe for supplying the 1995 eclipse image.

4 . References

1. Bolivia meeting, 1995, Proceedings of the International Symposium of the Total Solar Eclipse of November 3, 1995, edited by M.S.Raljevic, F.Zaratti, and J.M.Pasachoff.
2. Hiei,E., Inoue,K., Takahashi,N., 1995, The Solar Corona at the 1994 Eclipse, Rencontres du Vietnam, The Sun and Beyond, Ho Chi Minh City, Vietnam.
3. Hiei,E., Inoue,K., Takahashi,N., 1996, Observation of Total Solar Eclipse of 24 October 1995, in Kodaikanal Report, in press.
4. Kodaikanal Report, 1996, in press.
5. McIntosh.P.S, 1995, private communication.
6. Mouradian,Z., Prevot, G., Zlicaric,G., and Garcia A-D., 1995, Solar Phys.Letters.
7. Shiota,K, 1994, Sky and Telescope, November, p18- 20.
8. The Yohkoh Mission, 1991, Solar Phys., 136, 1-110.

DISTRIBUTION ON HALFWIDTHS FOR THE GREEN CORONAL LINE (5303Å)

I.V. ALEXEEVA, I.S. KIM
Sternberg State Astronomical Institute, Moscow State University
13, Universitetsky pr., 119899 Moscow, Russia

O.T. MATSUURA, E. PICAZZIO
Instituto Astronomico e Geofisico, Universidade de São Paulo
Av. Miguel Stefano, 4200, CEP 04301-904 - São Paulo, SP, Brazil

S. KOUTCHMY
Institut d'Astrophysique de Paris, CNRS
98 bis, bd Arago - 75014 Paris, France

Abstract.   Distribution on halfwidths for Fe~XIV (5303Å) coronal line is analyzed. The Fabry-Perot interferogram obtained during the total solar eclipse of November 3, 1994 is used. The profiles not distorted by the white light corona contributions as well as by ghost images were selected. A multimodality with probability of 87% is found in the histogram. Average values of halfwidths differ for different coronal structures (0.91Å for SW~helmet, 0.85Å for NW and 0.81Å for SE coronal regions). An evident increase of the halfwidths with distance is revealed for SW-helmet structure.

1. Introduction

Data on halfwidths (FWHM) obtained by non-eclipse coronagraphic and eclipse techniques strongly differ, varying from 0.70 Å to 1.4 Å [1, 2, 3, 4]). As a rule, a spectrograph or a Fabry-Perot interferometer (or an etalon) were used as a dispersing element by the above-mentioned authors. To clarify the question about the actual value of the halfwidth, a distribution of the green line emission at the halfwidth for the whole corona should be analyzed.

The Fabry-Perot experiment, aimed to receive the intensity, velocity, and half-width fields, was carried out in Criciuma (Brazil) during the eclipse of November 3, 1994. A coronagraph attached by a Fabry-Perot etalon (free spectral range = 4.35 ± 0.02 Å, FWHM = 0.23 Å), a narrow passband interference filter (FWHM = 6 Å, $\tau_{5303Å}$ = 21%), an image intensifier with a factor of 8, and an ORWO X-ray film (S(5303 Å) ≈ 650 ASA) were used. The instrumentation will be described somewhere else later. A set of exposures of 1, 3, 10, and 30 s revealed coronal spectra up to distances of 1.5 $R_\odot$. The reference spectrum of Fe XIV (5302.3 Å) was exposed on the same interferogram simultaneously with the coronal one. Absolute and relative intensity calibrations were made before and after the totality phase.

7

8

The summary of the first step of data analysis for the green line halfwidths deduced from the exposure of 10 s is presented below.

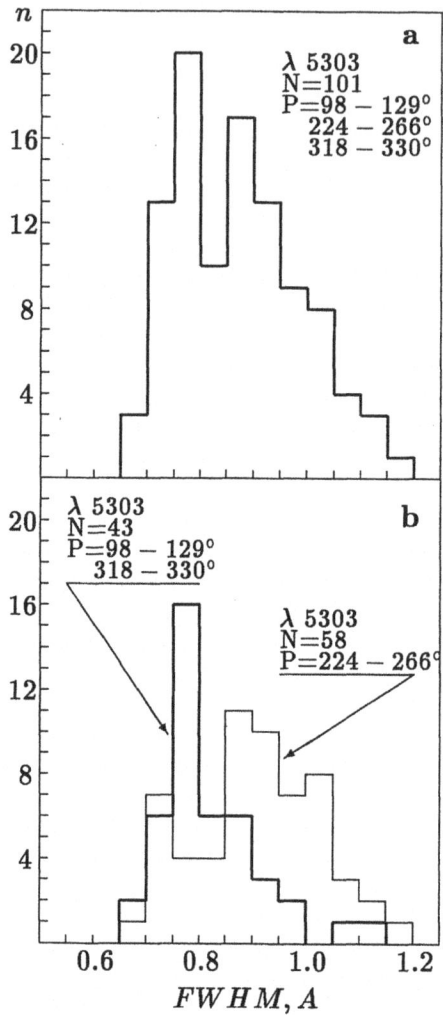

Figure 1. Distribution on the value of halfwidths: a - for the whole corona; b - for different coronal structures. N - total number of profiles

2. Photometry and Data Reduction

The ORWO X-ray films were developed by MWP-2 for 12 min at 20.5° C ($\gamma = 2.34 \pm 0.20$). The center of the fringe system was determined using the reference spectrum. Photometric radial cuts were made with the slit width corresponding to $7 \times 36''$ on the film. The accuracy of halfwidths determination varied from 0.03 to 0.05 and was mainly limited by the grain noise. 101 profiles not distorted by the white light corona structures and ghost images were selected for further analysis.

3. Results

Distribution on the value of halfwidths is shown in Fig. 1a for the whole corona. A multimodality is revealed on the histogram. The fiducial probability for minimum at 0.80--0.85 Å to exceed its half depth was calculated using Student's distribution and equaled 87%. This gives us an impetus to analyze separately the histograms for different white-light coronal structures. The distributions for the SW helmet structure (position angle P=224-266°), SE (P=98-129°) and NW (P=318-330°) coronal regions are shown in Fig. 1b. The average values of halfwidths (FWHM) for SW helmet structure, for NW and SE coronal regions are found to be equal 0.91 Å, 0.85 Å and 0.81 Å respectively. The averaged FWHM with the RMS error and the so-called "turbulent" velocities (V_t) are presented in Table 1, while T_i is (1.3 10^6) K being. Parts

Position angle	Table1 FWHM, Å	V_t, km/s
224-266°	0.91 ± 0.05	23.8
318-330°	0.85 ± 0.04	20.8
98-129°	0.81 ± 0.03	19.1

of two inter-ference fringes imposed on SW helmet (P=224-266°) covered the range of (1.2-1.5)$R_\odot$. The averaged values of FWHM found for (1.21÷1.33)$R_\odot$ and (1.43÷1.52)$R_\odot$ indicate an evident increase of the halfwidths with distance:

(0.86÷0.05) Å at (1.21÷1.33) $R_\odot$ and
(0.94÷0.05) Å at (1.43÷1.52) $R_\odot$.

4. Conclusions

Disagrements in the values of FWHM, cited by different authors, could be mainly explained by different coronal structures observed. The 3D structure and as a consequence the integration along the line-of-sight, seem to be responsible for the variation of the average value of the halfwidths.

In terms of "turbulent" velocities, the increase of the halfwidths with distance found for SW helmet structure is presented by the expression

$$V_t = \left(17 + 17 \frac{h}{R_o} \right) km / s$$

where h is the height of the point expressed in $R_\odot$.

Acknowledgements

The development of the instrumentation was supported by the grants N 93-02-2876 of the Russian Bound for Fundamental Researches (RBFR) and N 93-8-229 of the Russian Federal Program for Astronomy. Observations in Brazil were supported by the grants FAPESP 94/2643-3 (Fundacao de Amparo'a Pesquisa do Estado de Sao Paulo) and SNPq 452339/94-4 (Conselho Nacional de Desenvolvimento Cientifico e Tecnologco), data reduction by the grant N 96-02-17753 of RBFR. Our special thanks to Drs. O. Bougaenko and V. Brouevitch (Sternberg State Astronomical Institute) for their help in the preparation of the eclipse equipment and also for S. Faraco (Universidade Estadual de Londrina UEL) assistance in observations.

References

[1] Delone, A.B., Makarova, E.A. and Yakunina, G.V. (1988) Evidence for Moving features in the Corona from Emission line Profiles Observed during Eclipses, *J. Astrophys. Astron.*, **9**, 41-47

[2] Hirshberg, J.G., Wourtes, A. and Hazelton, L. (1971) Interferometric studies of spectral lines in the Solar corona, *Solar Phys.*,**21**, 448-451

[3] Liebenberg, D.H., Bessey, R.J. and Watson, B. (1975) Observed coronal temperatures at 1.37 $R_\odot$ in the region of a helmet structure, *Solar Phys.*, **40**, 387-396

[4] Makarova, V.V., Makarov, V.I. and Nikolsky, G.M. (1971) Profiles of 5303Å and 6374Å coronal lines based on observations by Great Coronagraph, *Astron. Zh*, **48**, N 2, 360-364

RESULTS OF POLARIZATION OBSERVATIONS OF THE WHITE-LIGHT CORONA

O.G. BADALYAN, M.A. LIVSHITS
Inst. of Terrestrial Magnetism, Ionosphere and Radio Wave Propagation
142092 Troitsk, Russia

J. SYKORA
Astronomical Institute, Slovak Academy of Sciences
05960 Tatranska, Lomnica, Slovak Republic

1. Introduction

Present paper deals with measurement and interpretation of the white-light polarization in the most pronounced coronal structures, i.e. in the helmet-like streamers and coronal holes. Observations were performed during three solar eclipses (1973, 1980, 1991), related to different phases of solar cycles. Reasons for the experiment, observations and their processing, and polarization of the global solar corona are described in detail by Badalyan, Livshits and Sykora [3, 4], and the references therein.

The eclipse images of the solar corona contain information covering large interval of heliocentric distances and substantially complement the space observations and the measurements of the ground-based coronagraphs. Coronal polarimetry is very efficient in identifying the discrete coronal structures and in studying physical conditions of them. Except of that, additionally to the two-dimensional images of the solar corona, our interpretation of the polarimetric measurements allows to estimate characteristic dimensions of the the majority of coronal structures along the line of sight and to understand, at least qualitatively, their geometry.

It should be perhaps mentioned that the space resolution of our measurements is rather high (5--10 arc seconds in the image plane) which, in principle, makes possible to study also far smaller coronal features than are the streamers and coronal holes. Presently achieved accurracy of our white-light corona polarization is not worse than 5% and, under a specially careful approach it can reach 3%.

2. Polarization in the Helmet-Like Streamers and Coronal Holes

The helmet-like streamers are the most outstanding large-scale features seen on the white-light coronal images. The degree of polarization in the streamers reaches up to

11

Z. Mouradian and M. Stavinschi (eds.),
Theoretical and Observational Problems Related to Solar Eclipses, 11-14.
©1997 *Kluwer Academic Publishers.*

12

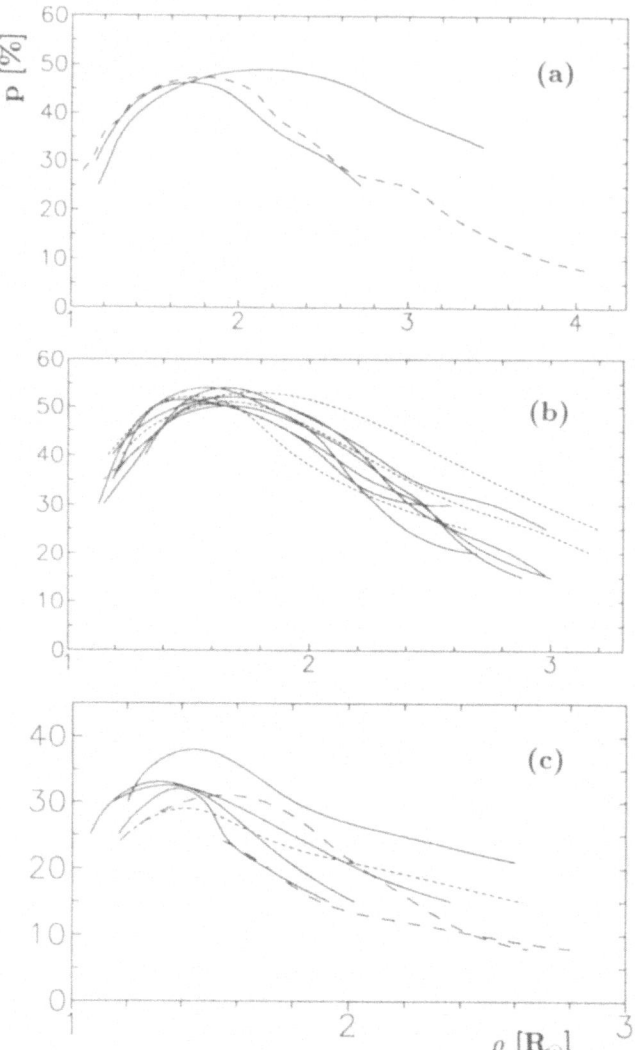

Figure 1. Radial distribution of the degree of polarization in the streamers and coronal holes:

(a) -- Eclipse of June 30, 1973, $P = 295°$ according to our measurement (shorter solid line), data of Koutchmy, Picat and Dantel, [7] (dashed line) and our measurement of July 11, 1991 at $P = 35°$(longer solid line).

(b) -- Eclipse of February 16, 1980 according to our data (solid lines) for position angles $P =$ (approximately) 50°, 70°, 130°, 203°, 235° and 315°, and according to Dürst, 1982 (dotted lines) for $P = 70°$, 203°, 315°. (See also [3]).

(c) -- Our measurement of the radial distribution of the degree of polarizationin four coronal holes: N-1973, SE-1973, SE-1991, S-1980 (the solid lines from above), the results of Koutchmy, Picat and Dantel [7]) for the holes:N-1973, SE-1973 (dashed lines from above) and that of Durst [6] for the hole S-1980 (dotted line).

47--55 % at distances of 1.6 -- 1.8 $R_\odot$. The highest degrees of polarization are found in the huge streamers at distances over 2 $R_\odot$.

Two upper panels of Figure 1 show dependence of the degree of polarization of heliocentric distance for several well-defined streamers. Figure 1a presents this dependence for the classical helmet streamer of the June 30, 1973 corona at the west limb (position angle P = 295°). Particularly for this feature our results are in good agreement with those of Koutchmy, Picat and Dantel [7]. In the same panel, the corresponding distribution of polarization for the huge NE-streamer of the 1991 eclipse is plotted according to our measurement. Comparison of the distributions of polarization of these two streamers shows that in the huge streamers the maximum of polarization is situated at higher levels than it is in more regular corona, and that the high values of p are maintained up to larger heliocentric distances.

Results showing the measured polarization in the streamers of the 1980 eclipse are compiled in Figure 1b according to Durst [6] and our measurements. Durst's data yield somewhat higher values of p, particularly at larger heights, but the differences remain within 5--10%. Irrespective of the relatively small streamers of the maximum 1980 corona, the distribution of polarization in some of them does not differ substantially from that found by us for the 1991 eclipse [3].

Theoretical considerations revealed that the values of polarization and brightness found in the streamers can be understood only under an assumption of a considerable concentration of coronal matter to the plane of the sky [3,2]. There are reasons to assume that the huge, predominantly equatorial streamers, are composed of several separated structures overlapping one another along the line of sight, with a background corona between them. Overlapping of such structures is, for example, clearly seen in the NE-quadrant of the July 11, 1991 eclipse.

Another conspicuous elements of the large-scale coronal structure are the coronal holes. In white-light radiation they are regions of remarkably reduced brightness of the (K+F)-corona, the maximum of which is located relatively low above the solar limb. The followings were observed during the three analysed eclipses: a large north-pole coronal hole CH and a smaller CH close to P = 145° in 1973 and several small CHs during the period of high solar activity in 1980 and 1991. Our measurements of four CHs are summarized in the lowest panel of Figure 1: N-1973, SE-1973, S-1980, SE-1991. Several measurements of other authors, related to the same holes, are added.

Figure 1c, as well as, the analysis of the SW-1970 hole P = 225° carried out by Durst [5], affirm that the maximum degree of polarization is nearly 30% at r= 1.3--1.5 $R_\odot$. At larger distances from the Sun, the polarization decreases rapidly, and above 3 $R_\odot$ the measurements become very difficult owing to the dominant role of the F-corona radiation. The model of the spherically symmetrical coronal atmosphere allows, unlike the streamers, both the distributions of brightness and polarization in the coronal hole to be explained simultaneously. Low degree of polarization,found in the coronal holes, is evidently connected with the known fact that the plasma densities are about 8--10 times lower than those in the streamers.

We have not found any substantial deviations of the direction of polarization from the tangential direction (electric vector of the electromagnetic radiation is considered), neither in the helmet-like streamers nor in the coronal holes. This is identical with most of the past investigations of orientation of the plane of polarization. We conclude that the direction of the white-light corona polarization does not seem to be sensitive to the weak magnetic fields present in the studied range of heliocentric distances.

3. Conclusion

Our study evidently confirms that the basic peculiarities, seen in the polarization of the solar corona, are connected with the streamers and coronal holes. It was proved that the degree of polarization is highly sensitive to these and any other structural features of the white-light corona. We have found promising that investigation of the coronal polarization gives a certain possibility of estimating the distribution of coronal plasma also in the direction perpendicular to the plane of the sky by evaluation the theoretically required plasma concentration towards this plane. This is particularly possible in the case of the well-developed helmet-like streamers.

Acknowledgements

This work was supported by the grant No. 96-02-18383 of the Russian Foundation of Basic Research and by the VEGA grants Nos. 02/2004/95 and 02/2007/95 of the Slovak Academy of Sciences.

References

1. Badalyan, O.G. (1995) *Transactions Astron. Astrophys.* **9**, 205.
2. Badalyan, O.G., Livshits, M.A.,(1994) in *Solar Coronal Structures* ,V. Rusin, P. Heinzel and J.-C. Vial (eds.), Veda Publ. House, Bratislava, 77.
3. Badalyan, O.G., Livshits, M.A., Sykora, J. (1993) *Solar Phys.* **145**, 279.
4. Badalyan, O.G., Livshits, M.A., Sykora, J. (1996) *Solar Phys.*,(in press)
5. Dürst, J.(1976), *Solar Phys.* **50**, 457.
6. Dürst, J.(1982), *Astron. Astrophys.* **112**, 241.
7. Koutchmy, S., Picat, J.P., Dantel, M.(1977), *Astron. Astrophys.* **59**, 349.

DEGREE POLARIZATION DISTRIBUTION IN THE SOLAR CORONA OF 03 NOVEMBER 1994.

I.A. BELENKO, O.I. BOUGAENKO, I.S. KIM
*Sternberg State Astronomical Institute. Moscow State University,
13, Universitetsky pr., 119899 Moscow, Russia*

S. KOUTCHMY
*Institut d'Astrophysique de Paris, CNRS,
98 bis, bd Arago - 75014 Paris, France*

O.T. MATSUURA, E. PICAZZIO
*Instituto Astronomico e Geofisico Universidade de São Paulo
Av. Miguel Stefano, 4200 CEP 04301-904, São Paulo, SP, Brasil*

Abstract Preliminary data reduction of the color index polarization wave experiment carried out in Brazil (Chapeco town) is described. Distributions of polarization degree (polarization images) in the range 1.1-1.6 $R_\odot$ are obtained by an algorithm developed on the basis of the Stokes vector presentation and IDL software. The polarization image of the "quasi-white-light" corona (summary of the blue and red spectral region emissions) is analyzed. On the average, no fitt is found with the van de Hülst model for the inner corona at heliographic latitudes of ±45°. The polarization degree reaches a plateau at 1.2-1.35 $R_\odot$.

1. Introduction

A color index polarization wave experiment was carried out in Brazil (Chapeco town) during the total solar eclipse of 03 November 1994. The experiment was aimed at studying the polarization within the innermost corona for any spatial and time variations, differences between the blue and red spectral intervals, or van de Hülst model fit, among other possibilities. A portable coronagraph-polarimeter (with an entrance aperture of 55 mm and the equivalent focal length of 339 mm, a rotating polarizer synchronized with a motion-picture professional camera and placed in the plane of the Lyot stop) has been developed and manufactured at Sternberg State Astronomical Institute. The camera films at a rate of five frames per second, and a half-turn of the polarizer corresponds to 12 consecutive frames. Details concerning the instrumentation are given by Kim et al. [3]. A 35 mm color negative film (45 ASA) was used. 1250 frames were obtained during the total eclipse of 244 s. Absolute calibration was based on the solar disk exposures.

The results of the first step of data reduction of 24 frames taken through the green filter (λ_{max}=533nm), centered on the totaleclipse are presented below. The film was digitized to a pixel size corresponding to 6"× 6". The moving average smoothing

15

*Z. Mouradian and M. Stavinschi (eds.),
Theoretical and Observational Problems Related to Solar Eclipses, 15-19.
©1997 Kluwer Academic Publishers.*

resulted in the angular resolution of 20" for densities of 0.3 and more, while the accuracy is 1-2%. Accuracies of 2-3% are achieved for densities of 0.15-0.3 and angular resolutions of 30". For low densities (< 0.15), an accuracy of 3-4% is achieved for angular resolutions of 40" ÷ 66".

2. Observations and Data Reduction

Data reduction is based on IDL software. After centering the 24 frames together, the averaged-density corona image $\mathbf{D} \equiv D(i,j)$ is created, where i,j are rectangular integer coordinates of pixel images referenced to the center of the Sun. The density distribution in each frame is also used as a noise level indicator in generating the coronal polarization stastistics. S_k-intensity frames are developed from the density frames, containing information about the polarization parameters distribution in the corona image, which could be extracted by statistical processing.

According to Malus law, the partially linear-polarized emission intensity after passing through an ideal polarizer is $S \sim I + Q\cos 2\alpha + U\sin 2\alpha$, where (I,Q,U) are the first three components of the Stokes vector for every image point and α is the position of the polarizer polarization plane in the (Q, U) coordinate system. Below, the I, Q and U intensities are considered to depend on (i,j) coordinates, i.e. frames $\mathbf{I} \equiv I(i,j)$, $\mathbf{Q} \equiv Q(i,j)$, and $\mathbf{U} \equiv U(i,j)$.

For the experiment, the polarization plane was shifted by a constant value $2\pi/24 = 15°$ from one frame to the next, so that for the kth frame $\alpha_k = \alpha_0 + \pi k/12$, where α_0 is the position of the polarization plane in the first frame. Thus, 24 equations can be written for every (i,j) pixel using the 24 consecutive frames:

$$S_k \equiv S_k(i:j) = \mathbf{I} + \mathbf{Q}' \cos \frac{\pi k}{12} + \mathbf{U}' \sin \frac{\pi k}{12}, \quad k=0,1,...,23,$$

where vector $(\mathbf{Q}'\ \mathbf{U}')$ is related to (Q,U) by $2\alpha_0$ rotations unknown beforehand with enough accuracy.

The solution of the system $\{S_k\}$ relative to $\mathbf{I}$, $\mathbf{Q}'$ and $\mathbf{U}'$ is based on the least squares method applied to:

$$\mathbf{I} = \frac{1}{24} \sum_{k=0}^{23} S_k, \quad \mathbf{Q}' = \frac{1}{24} \sum_{k=0}^{23} S_k \cos \frac{\pi k}{12}, \quad \mathbf{U}' = \frac{1}{24} \sum_{k=0}^{23} S_k \sin \frac{\pi k}{12}.$$

Because of the high coronal brightness gradient, any further statistical processing, for spatial smoothing, is more effective with $\mathbf{q}' \equiv q'(i,j) = \mathbf{Q}'/\mathbf{I}$ and $\mathbf{u}' \equiv u'(i,j) = \mathbf{U}'/\mathbf{I}$. This is especially important in the low density regions, where the polarization noise reaches ± 100 %.

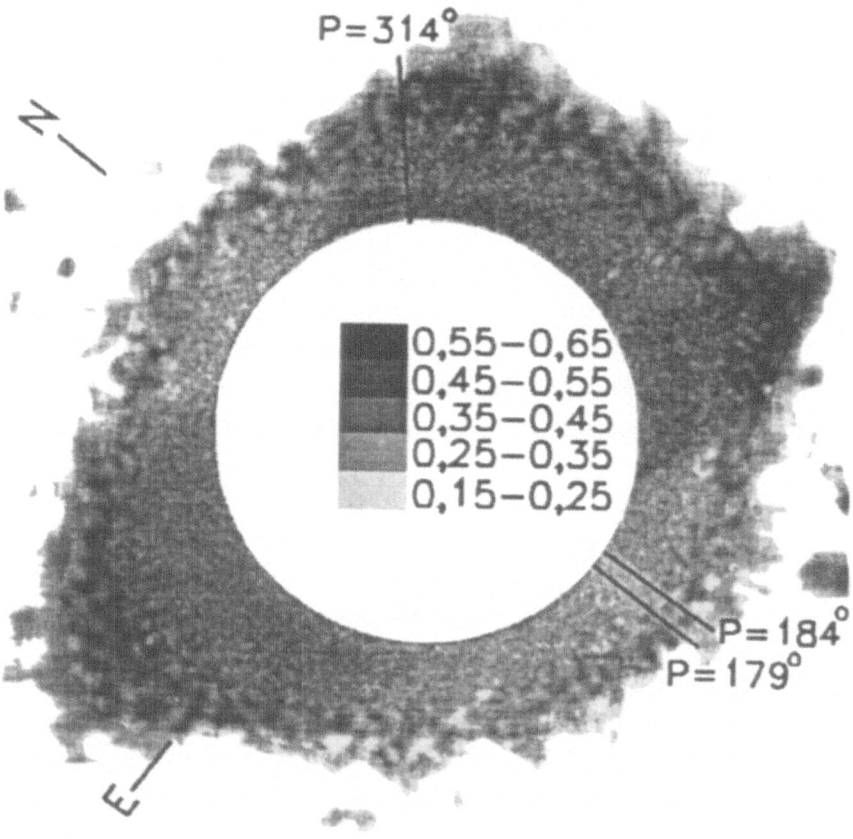

Figure 1.

Thus, **q'** and **u'** are the components with alternating signs of polarization degree, and the modulus of polarization degree equals $p = \sqrt{q'^2 + u'^2}$. On the other hand with the **Q'** and **U'** gradients, the **q'** and **u'** gradients are comparatively low according to van de Hülst's model [2] confirmed by different experiments [1, 5, 6]. It should be noted that any attempt to smooth the noisy p-frames always resulted in an overestimated polarization degree. In general, coronal emission is polarized due to electrons and small dispersed aerosol scattering [8]. In the ideal case when every point of the image was converted to local Cartesian coordinates with the axis always directed to the Sun's center, the radial component of the polarization vector was zero, while the tangential component coincided with the polarization vector modulus. This did not hold in regions of anomalous polarization [7]. The conversion to a local coordinate system for the vector of polarization the (**q'**, **u'**)-vector is rotated by an angle $2\psi \equiv 2\psi$ (i,j) determined as

$$\cos 2\psi = (i^2 - j^2)/\sqrt{i^2 + j^2}; \sin 2\psi = 2ij/\sqrt{i^2 + j^2}.$$

18

The tangential **t'**- and radial **r'**-frames are created for the rectangular projections **q'** and **u'** by such a transformation

$$
\begin{array}{cc}
\mathbf{t'} & \cos 2\psi \quad \sin 2\psi & \mathbf{q'} \\
\mathbf{r'} & -\sin 2\psi \quad \cos 2\psi & \mathbf{u'}
\end{array}
$$

The angle α_0 is determined by the expression $\tan 2\alpha_0 = \langle\, \mathbf{r'}\, \rangle / \langle\, \mathbf{t'}\, \rangle$ assuming that the average radial component $\langle\, \mathbf{r}\, \rangle$ is equal to zero.

The rotation by $2\alpha_0$ results in conversion of the (**t'**,**r'**)-frames into the (**t**, **r**)-frames with a mean value of zero **r**. Only those regions of the images with densities of $0.2 \div 0.8$ in the averaged **D**-frame were used to calculate the mean $\langle\, \mathbf{r'}\, \rangle$ and $\langle\, \mathbf{t'}\, \rangle$. The statistical processing algorithm is presented in detail by Kim et al. [4].

3. Results

Figure 1 shows a computer half-tint polarization image of the white-light (summary of the blue and red emission) corona. It looks similar to a conventional coronal image revealing evident coronal features. On the average, "polarization brightness" is minimum near the limb, increases rapidly to a certain level with distance and is practically constant from 1.2 to 1.4 $R_\odot$, then increases again. The plumes structures are practically invisible in the image. It should be noted for outer coronal regions (densities < 0.06) that there are systematic polarization degree errors of 5%.

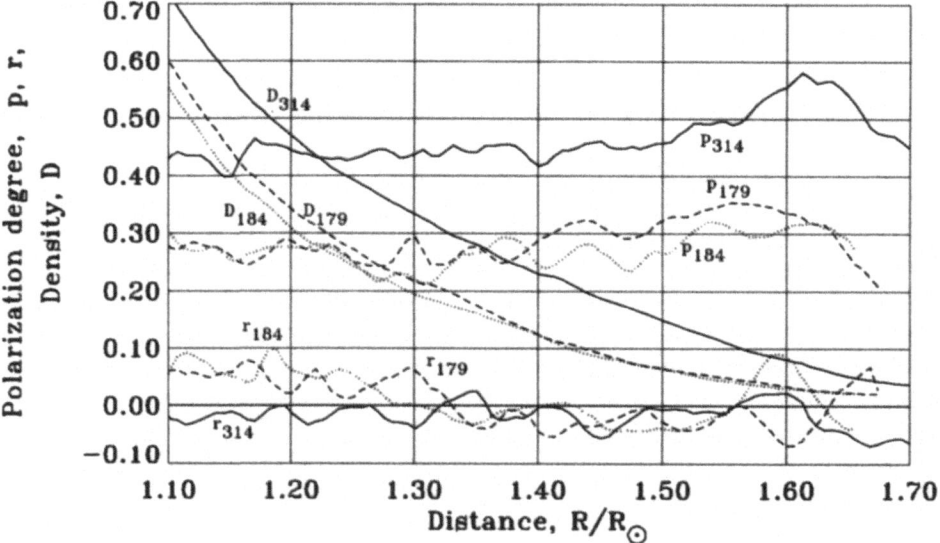

Figure 2.

Three radial scans are shown in Figure 1. The first, at a position (P) of 314°, corresponds to the NW streamer where the highest polarization degree (55%) is found. Two other scans correspond to the south polar region: within the plume $P=179°$ and between the plumes $P=184°$.

The dependencies of density (**D**),polarization degree (**p**), and the radial projection (**r**) of polarization degree versus distance are shown in Figure 2. The **D**-curves correspond to the 24-frame average density. In case of Thomson scattering, the existence of a radial component represented by **r**-curves is caused by different errors of measurements. In general, variations of **r**-curves characterize the actual accuracy of the polarization degree determination.

4. Conclusions

1. In the range 1.1 - 1.7 $R_\odot$, at heliographic latitudes of ±45°, there is monotonoic increasing in the polarization degree corresponding to the van de Hülst model. Polarization degree is practically constant in the 1.2 - 1.4 $R_\odot$ range, and differs by a factor of 1.5 for different coronal structures. For example, the averaged polarization degree equals 44% for equatorial regions and 28% for polar. Most likely, the above-mentioned plateau is caused by the 3D structure.
2. No remarkable difference is found in the polarization degree between plumes and the background corona.

Acknowledgments

The development and manufacture of the instrumentation were supported by the grants N93-02-2876, of the Russian Fund for Fundamental Research (RBFR), and N93-8-229 of the Russian Federal Program for Astronomy, observations by Brazilian grants CNPq 452422/94-9 (Conselho Nacional de Desenvolvimento Cientifico e Tecnologico) and Of.PRP 191/17094 (Pro'-Reitoria de Pesquisa da Universidade de Sao Paulo), and data reduction by grant N 96-02017753 (RBFR).

References

[1] Clette, F. and Gabril, J.R. (1992) The July 11, 1991 solar eclipse observed in polarized white light, *Proc. 1st SOHO Workshop* ESA SP--**348**, 351--353
[2] van de Hülst, H.C. and Bull, Y.C. (1950) *Astron. Inst. Netherl.* **11**, 135
[3] Kim, I.S., Bougaenko, O.I., Belenko, I.A., Guetman, F.I., and Koutchmy,S. (1997) Polarization observations of the white-light solar corona of 03 November 1994, *Solnechnye Dannye* (in press)
[4] Kim, I.S., Bougaenko, O.I., Belenko, I.A. and Matsuura, O.T. (1996) Coronagraph-polarimeter. The algorithm for the Solar corona images creating, *Radiophisika* **12** (in press)
[5] Kishonkov, A.K. (1974) Measuring of the Solar corona polarization degree and directions for the solar eclipse 10 June 1972, *Solnechnye Dannye* **8**, 63--69
[6] Koutchmy, S., Molodensky, M.M., Nikolsky, G.M., Filipov, B.P. (1993) On the Measuring Solar Corona Polarization, *Astron. Zh* **70**, 563-570.
[7] Nikolsky, G.M., Sazanov, A.A. and Kishonkov A.K. (1977) *Solar physics* **53**, 76-79
[8] Shklovsky, I.S. (1962) *Physics of the Solar Corona*, FM, Moscow, 152.

FUNDAMENTAL CONTRIBUTIONS OF THE SOLAR ECLIPSE OBSERVATIONS TO SOLAR PHYSICS IN THE SECOND HALF OF THE 20TH CENTURY

RUDOLF A. GULYAEV

IZMIRAN, 142092 Troitsk, Moscow Region, Russia

Abstract. Astrophysical observations of solar eclipses began in the middle of the 19th century. The most important observations for the 100 years from 1851 to 1952 have been listed in well-known compilations by S.A. Mitchell, complemented by H.C. van de Hulst. The second half of the 20th century is also marked by a number of important scientific findings from eclipse observations. We are compiling a review of major observations, extending the list by Mitchell up to the end of the 20th century. In this paper two results of particular concern to the author are discussed:
(1) Spectral observations during the eclipses of 1970 and 1973 showned that coronal material penetrates very deeply into the chromosphere;
(2) The eclipse of 1991 yielded direct observational evidence identifying a coronal streamer belt and the base of the heliospheric current sheet.

1. Introduction

Astrophysical observations of solar eclipses began in the middle of the 19th century. When photography and spectral analysis were applied to eclipse triggered an avalanch of important discoveries.

By the beginning of the 20th century, a most of the cream had been skimmed off this new flood of observational data. A detailed list of the most important observations for the 100 years from 1851 to 1952 can be found in the well-known reviews by Mitchell [1], among others. Nevertheless, a number of subsequent eclipses have also yielded major scientific findings. To continue the list by Mitchell, we have compiled a review of important observations made during the second half of the current century. This new list is rather long, so we will present it elsewhere. Here, we will discuss only two results of particular concern to the author .

2. The Solar Corona at Very Low Heights

In the middle of the 20th century the solar atmosphere was imagined to be structured general by as follows [2]: the chromosphere ranges up to height of about 10,000 km, the corona begins at about 20,000 km, and an extensive transition layer lies in between. In its most complete form, this concept was the model of Ivanov-Kholodny and Nikolsky [3].

At the eclipse of 1952, the High Altitude Observatory expedition succeeded in obtaining a sequence of slitless spectrograms at intervals of 0.4 sec, giving a height

21

Z. Mouradian and M. Stavinschi (eds.),
Theoretical and Observational Problems Related to Solar Eclipses, 21-24.
©1997 *Kluwer Academic Publishers.*

resolution of 110 km. Using these spectrograms, Athay and Roberts [4] studied a distribution of integrated coronal lines intensities '(Fe XIV λ 5303 and Fe XI λ 7892) in the height range from 3000 km to 46,000 km. A sensational discovery was made for the λ 7892 line: the surface brightness of the line is maximum at a height of less than 10,000 km. Athay and Roberts concluded that coronal line emission begins at the very top of the chromosphere and maybe even in between the spicules. Many investigators thought this result was accidental.

At the eclipse of 1970, Japanese observers [5] set up a special experiment to study the behaviour of the coronal lines λ 5303 and λ 6374 very close to the solar limb. For two regions referring to the quiet chromosphere, they derived the integrated intensity distribution of these lines in the height range from 800 km to 10,000 km. Figure 1 (left) shows results for one of the regions. It can be seen that the integrated intensities of the lines increases down to about 1000 km - especially the red line,which shows a marked intensification below 3000 km. The green line data are fitted by a single exponential term; the red line can be represented suitably by a sum of two exponential terms. Giovanelli [6] raised some doubts in connection with the conclusions of Kanno et al. So Kanno [7] made further analysis of intensity distributions for the coronal lines using new observational data from the eclipse of 1973 and confirmed previous conclusions. According to observations made during that eclipse, the exponential increase of the λ 6374 line intensity certainly continues down to $h \approx 1000$ km [8].

The author of the present paper made an independent study of the coronal line behaviour at very low heights. At the total solar eclipse of 1973, we have carried out observations of slitless coronal spectrum in the region of the line λ 6374 [9]. Fifty-three spectrograms were obtained. Seventee regions were selected for the photometric

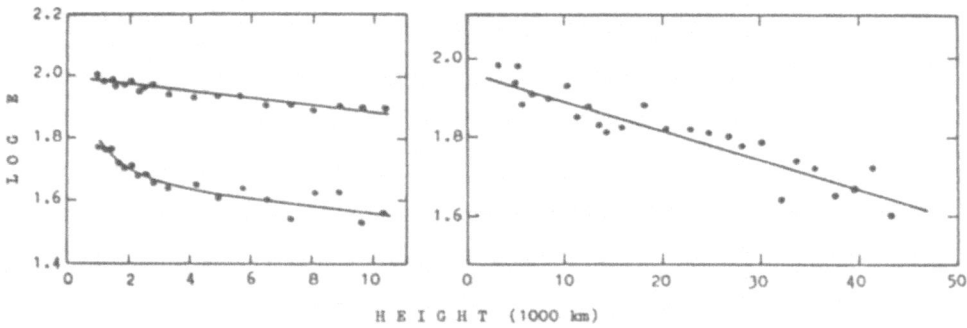

Figure. 1. Left: Integrated intensities of the Fe XIV and Fe X lines according to Kanno et al. [5]. Right: Integrated intensity of the Fe X line according to Gulyaev and Ajmanova [9]. *E* in relative units.

analysis: 11 at the eastern limb and 6 at the western. The regions were chosen so as to include the diversity of brightnesses, activity levels and structural properties. The radial variation of the integrated intensity of λ 6374 line were obtained for the heights ranging from about 3000 km to about 50,000 km in 17 regions distributed around nearly the whole limb. The conclusions drawn by the Japanese observers are completely confirmed: in all cases, the intensity of the λ 6374 line increases exponentially towards the limb down to at least 3000 km. For example, Figure 1 (right) represents the graph of lg $E = f(h)$ for the region with the position angle of 73°.

So the exponential increase of the red line intensity down to at least 3000 km alwaysholds, independent of the heliographic latitude, absolute brightness, or morphological properties of a particular region at the limb. Thus it was stated that we are dealing with a rather general character of the solar atmosphere: coronal material fills the space everywhere between spicules down to the very low chromospheric layers. At presentthis picture is generally accepted.

3. The Outer Solar Corona as the Base of the Heliospheric Current Sheet

The total eclipse of July 11, 1991 was prominent in stimulating studies concerned with the structure and evolution of the outer solar corona. The eclipse occurred soon after the peak of sunspot cycle 22, so the corona should have had a shape close to the "maximum" type, as was ordinarely thought. But the actual corona of 1991 looked completely different. It was quite similar to the minimum type, although in contrast to the classical minimum corona it was very steeply inclined against the solar equator [10, 11] . So the corona of 1991 demonstrates the failure of the accepted pattern of coronal shape variation during the sunspot cycle.

The primary structural elements of the outer solar coron are known to be helmet streamers determining the global large-scale image of the corona on the whole. Many recent investigations have shown that helmet streamers are not distributed randomly above the solar surface, but instead, form a closed belt around the Sun, and this belt is the base of the heliospheric current sheet (HCS). The total solar eclipse of July 11, 1991 provided us with a fortunate chance to correlate the coronal structure directly with the HCS orientation. It was found that the orientation of the coronal streamer belt is indeed determined by the position of the magnetic dipole equator viz. by the HCS orientation [10-12]. The following eclipses of 1994 and 1995 confirmed these inferences concerning the blood relationship between the outer solar corona and the heliospheric current sheet [13].

Thus, observations of solar eclipses have provided direct experimental proof of validity of concept that the coronal streamer belt is a visible manifestation of the heliospheric sheet, and individual helmet streamers are details of the HCS structure. It follows from above that an observed coronal shape is conditioned first of all by orientation of the solar magnetic equator with respect to the Earth [14-16]. On July 11, 1991, the solar corona was very steeply inclined against the solar equator and was turned almost edge-on towards the Earth. Clearly, if the eclipse had occurred a week before or after 11 July, the coronal streamer belt wouldhave be seen nearly in the plane of the sky, and we would have observed a typical maximum corona.

The above considerations have an attractive consequence. Solar astronomers have long dreamt of looking at the minimum corona "from above",i.e. from the Sun's pole. Unfortunately, the spacecraft *Ulysses* is not carrying a coronagraph, so our chance to see the corona "from above" will not this time. But we know today that it is possible to see the corona from the pole side while remaining on Earth. Of course, what we mean by this is the solar magnetic poles and not the rotation poles. When the solar magnetic pole is located near the Sun's disk center, the coronal streamer belt is close to the plane of the sky, i.e.is seen face-on.

It is clear that a very steep inclination of the coronal mean plane against the solar equator is needed in order for such a caseto exist. The soonest opportunity for this is expected after 1998. Just after that time, the total solar eclipse of August 11, 1999 will occur. There is a very high probability that the solar corona of August 1999 will be rather flattened and steeply inclined against the solar equator plane. If its orientation, is edge-on we will again see the corona as it was observed in 1991. But it may just happenthat we will even luckier, and see the corona face-on i.e. from the magnetic pole side.

References

1. Mitchell, S.A. (1951). *Eclipses of the Sun*, 5th edition, Columbia Univ. Press.
2. Van de Hulst, H.C. (1953). In: G.P.Kuiper (ed.), *The Sun*, Univ. of Chicago Press,. 207-321.
3. Ivanov-Kholodny, G.S. and Nikolsky, G.M. (1962). *Astron. Zh.*, **39**, 777-791.
4. Athay, R.G. and Roberts, W.O. (1955). *Astrophys. J.*, **121**, 231-240.
5. Kanno, M., Tsubaki, T, and Kurokawa, H. (1971). *Solar Phys.*, **21**, 314-324.
6. Giovanelly, R. (1975). *Solar Phys.*, **43**, 377-379.
7. Kanno, M. (1975). *Solar Phys.*, **43**, 381-384.
8. Hanaoka, Y., Kanno, M., Kurokawa, H., and Tsubaki, T. (1986). *Solar Phys.*, **106**, 95-106.
9. Gulyaev, R.A. and Ajmanova, G.K. (1978). *Solnechnye Dannye*, **11**, 80-86.
10. Gulyaev, R.A. and Filippov, B.P. (1992). *Doklady Akademii Nauk*, **322**, 268-271.
11. Gulyaev, R.A. (1994). *Adv. Space Res.*, **14**, No. 4, 53-56.
12. Saito, T., Akasofu, S.-I., Kozuka, Y., Takahashi, T., and Numazawa, S., (1993). *J. Geophys. Res.*, **98**, No. A4, 5639-5644.
13. Gulyaev, R.A. and Molodensky, M.M. (1995). *Rev. Acad. Nac. Cienc. Bolivia,.* **69**, 11-17.
14. Gulyaev, R.A. (1992). *Solar Phys.*, **142**, 213-216.
15. Gulyaev, R.A. (1996). *Astron. and Astrophys. Transactions*, in press.
16. Sykora, J., Pinter, T., and Ambroz, P. (1995). *Rev. Acad. Nac. Cienc. Bolivia*, No. 69, 23-36.

POLARIZATION OF THE 530.3 nm CORONAL LINE AS OBSERVED ON JULY 11, 1991

O.G. BADALYAN
Inst. of Terrestrial Magnetism, Ionosphere and Radio Wave Propagation
142092 Troitsk, Russia

J. SYKORA
Astronomical Institute, Slovak Academy of Sciences
05960 Tatranska Lomnica, Slovak Republic

Observations of the FeXIV 530.3 nm emission coronal line permit to investigate physical conditions of the inner corona.

Using intensity and polarization measurements of this "green-line" we can obtain distributions of temperature and density, mainly in coronal loops (representing a

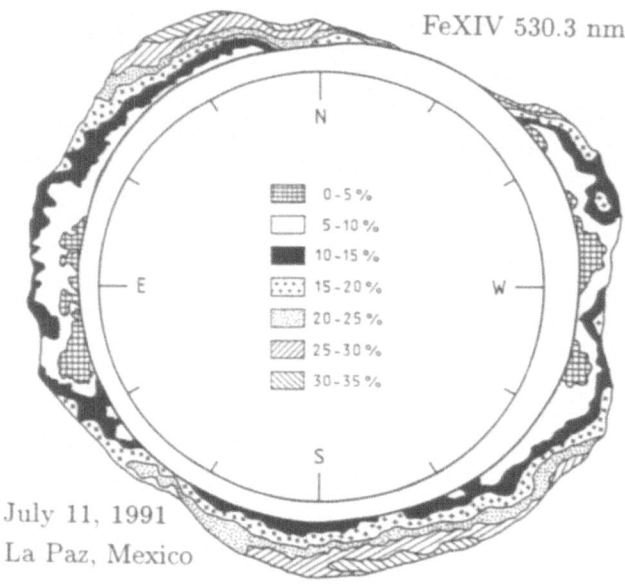

FeXIV 530.3 nm

▦	0 - 5 %
☐	5 - 10 %
■	10 - 15 %
⋯	15 - 20 %
▨	20 - 25 %
▨	25 - 30 %
▨	30 - 35 %

July 11, 1991
La Paz, Mexico

Figure. 1. Map of the green-line corona polarization. Inner circle represents the Sun, while outer circle corresponds to an "envelope" created from the moon's positions during four exposures.

Z. Mouradian and M. Stavinschi (eds.),
Theoretical and Observational Problems Related to Solar Eclipses, 25-29.
©1997 *Kluwer Academic Publishers.*

hot component of the solar corona). Polarization data contain also a valuable information about magnetic fields of the inner corona, particularly, they indicate direction of the magnetic lines of force. On the other hand, measurement of the white-light corona brightness gives us information about both the components of coronal plasma - hot and cold (background) ones.

Present observation was carried out by one of the authors (J.S.) in La Paz (Mexico) during the 1991 total solar eclipse under very good weather conditions. A 0.17 nm narrow-band interference filter was used to separate radiation of the 530.3 nm emission line. Altogether, four images (30 s exposures) of the green-line corona at different positions of a polarizer were taken. Except of that, the same procedure was used in obtaining three sets of the white-light corona polarized images (exposures 1/125 s, 1/15 s and 1 s). All the photographs are of excellent quality.

The white-light corona of the July 11, 1991 eclipse showed to be quite atypical for the actual phase of the solar cycle: highly flattened with the huge streamers situated at high latitudes near the N and S poles.

Image processing was carried out by using the microdensitometer of the Ondrejov Observatory. The fitting of all the images was performed in the Institute of the Problems of Information Transmission (Moscow). It was necessary to be sure that the same point of the corona was extracted from the corresponding four images for subsequent calculations. Resolution of about 5 arc seconds was maintained throughout the whole data analysis.

Experiment, observation, processing of the images and equations for calculation of coronal intensity and polarization are described in more details in Badalyan and Sykora [1].

As a result, we obtained a certain map showing distribution of the degree of the green-line polarization (Figure 1). As far as we know, such a map, displaying

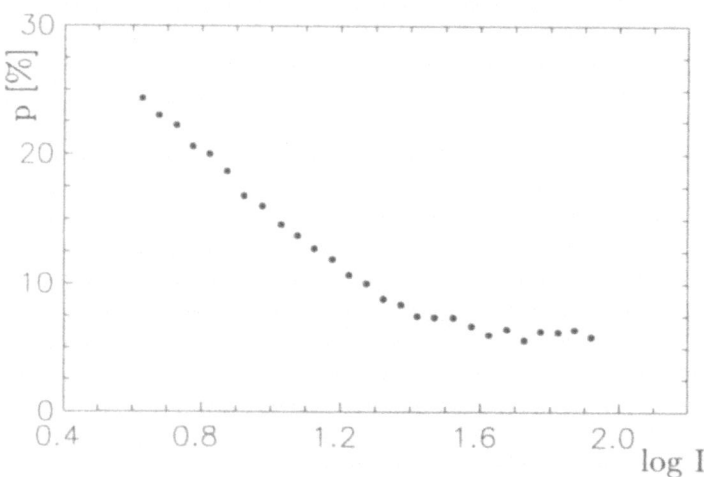

Figure 2. Average anticorrelation between the polarization and intensity of the green-line corona

polarization of the whole emission corona, is presented for the first time in history of the eclipse observations. One can see that the degree of polarization varies from 0 % to about 30 %. It increases with distance from the limb at all position angles. There is also a difference in the polarization rate between equatorial regions and regions of thehigh-latitude streamers. In the brightest green-line regions the polarization is low, while in the fainter regions it is high. This fact indicates certain anticorrelation between the polarization and intensity of the green line corona. Averaged course of this anticorrelation was derived from the entire matrix of data (see Figure 2).

Calculations revealed that the direction of polarization (electric vector) is more or less tangential to the solar limb in the high-latitude streamers, whereas, it deviate from the tangential direction by various angles in the equatorial regions. This may indicate an influence of the magnetic field.

We compared our results with those of other authors. In Figure 3 thefour sets of results are presented, including our data. Measurements of Arnaud, Hyder and Picat et al. were taken from Figure 7 of Picat et al. ([2] and the references therein). The upper curves refer to streamers, the lower ones to bright (active) regions (Picat et al. report only the curve for bright regions). We remark that the results of Picat et al. and those obtained by us are represented here also by the curves of mean polarization. It follows from Figure 3 that:

-- Each of the curves show an increase of the degree of polarization with distance from the limb;

-- All the four sets of results indicate polarization in the streamers to be higher in comparison with that in the coronal condensations (usually situated above the photospheric active regions);

-- We give somewhat higher degrees of polarization in comparison with the results of other authors. At the same time, we are most close to the measurement by Picat et al., relatied to the June 30, 1973 eclipse.

What information about the hot component of the inner corona could we get from the green-line polarization data? The upper level of this line is exited by two processes: electron collisions and photospheric radiation (photo-excitation). Correspondingly, the green-line intensity consists of two components:

$$I_\lambda = I_e + I_{ph}.$$

Let us denote:

$$D = I_{ph} / I_e.$$

However, only I_{ph} is polarized, while, $I_{ph} = I_t + I_r$, where I_t and I_r are tangentially and radially polarized components. Then, the observed polarization p is:

$$p = (I_t - I_r) / (I_t + I_r + I_e) = p_{ph} / (1 + 1/D)$$

where

$$p_{ph} = (I_t - I_r) / (I_t + I_r)$$

is polarization of the "photo-excited" component only. Consequently, we have:

$$I_{ph} = (p / p_{ph}) I_\lambda$$

$$I_e = (1 - p / p_{ph}) I_\lambda$$

Thus, in principle, these two components can be derived and, subsequently, some conclusions on physical conditions and processes in the corona can be deduced. The value p_{ph} can be calculated theoretically if a certain model of temperature and density distribution is assumed. In absence of the magnetic field, the value of p_{phj} is equal to the K-corona polarization, because the indicatrisse of a linear emission is the same as in case of the Thomson's scattering. In fact, the p_{ph} is reduced by the presence of the magnetic field. Therefore, a lower polarization found in the active regions could indicate a real presence and influence of the magnetic field.

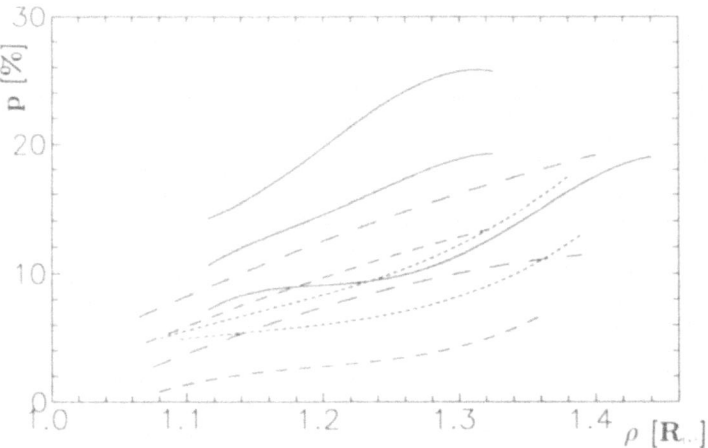

Figure 3. Degree of polarization as a function of the heliocentric distance. Solid lines - our data (low-equatorial zone, upper - regions of the high-latitudes streamers, middle - mean polarization), dotted lines-data of Hyder, dashed lines (short dashes) - data of Arnaud, dashed lines(long dashes) - data of Picat et al. [12].

After obtaining I_{ph} and I_e the distribution of coronal density can be calculated. At last, the cold (background) and hot (arising in hot loops) density components can be separated in the white-light radiation of the inner corona. Therefore, our experiment and procedure allow, in fact, the complete diagnostic of the coronal plasma to be carried out.

This work was supported by the grant No. 96-02-18383 of the Russian Foundation of Basic Research and by the VEGA grants Nos. 02/2004/95 and 02/2007/95 of the Slovak Academy of Sciences.

References

1. Badalyan O.G. and Sykora J. (1996) *Astron. Astrophys.*, (in press).
2. Picat J.P., Felenbok P., and Fort B.: (1979) *Astron. Astrophys.* **75**, 176.

MEASUREMENT OF THE CORONAL ELECTRON TEMPERATURE AT THE TOTAL SOLAR ECLIPSE ON 1994 NOVEMBER 3

K.ICHIMOTO, K.KUMAGAI, I.SANO, T.KOBIKI, T.SAKURAI
National Astronomical Observatory Mitaka, Tokyo 181, JAPAN

A.MUNOZ
Asociacion Chilena de Astronomia y Astronautica,
Marcoleta 485 - Of. H. Santiago, CHILE

Abstract. The coronal electron temperature was determined for the first time from the shape of weak depressions at 3900Å and 4300Åai in the continuous coronal spectra.It was found that the coronal streamer has overall temperatures of about 1.5-1.7 MK, while the coronal hole has evidently lower temperature of 0.9-1.1 MK.No significant height variation of the temperature was found in the streamer, while the coronal hole shows evidence of increasing temperature with height. An acceleration of the expanding motion between $1.1\,R_\odot$ and $2.0\,R_\odot$ by ~ 80 km s^{-1} was found in the streamer.

The shape of the continuous coronal spectra which is created by Thomson scattering of the photospheric light is a function of the thermal motion of the free electrons in the corona and thus provides a direct information on the coronal electron temperature (Cram, [1]). A spectroscopic observation was carried out at the total solar eclipse on November 3, 1994 in Putre, northern Chile, in order to investigate the electron temperature and the expanding motion of the corona (Ichimoto et al. [2]). The spectral images covering the wavelength of 3500Å ~ 4720Å were obtained with a 28-cm Cassegrain telescope and a small spectrograph with a 512x512 cooled CCD. The field of view was about 42 arc min. along the slit. Five exposures were obtained during the totality of 3-minutes for 3 different locations of the slit including the eastern streamer and the southern coronal hole. A spectrum of the uniform sky was taken before and after the eclipse to get the flat field in the slit direction and to guess the spectral distributions of the F-coronal component. Thin clouds covering the sky at the time of the totality caused a scattered light and introduced some uncertainties on the interpretation of the data for the faint part. However the shape of the spectra can be considered not to be affected seriously.

After correcting the sensitivity of the instrument and removing the F- and E-component from the observed spectra, we compared the continuous component of the spectra with the theoretical one which is calculated by assuming a coronal density model and the temperatures. Figure 1 shows examples of the observed continuous spectra and the calculated spectra for the spherical coronal model with the coronal temperatures of 0.6, 1.2, 1.8, 2.4, and 3.0 MK. The best fitting of the curves gives the

Z. Mouradian and M. Stavinschi (eds.),
Theoretical and Observational Problems Related to Solar Eclipses, 31-34.
©1997 *Kluwer Academic Publishers.*

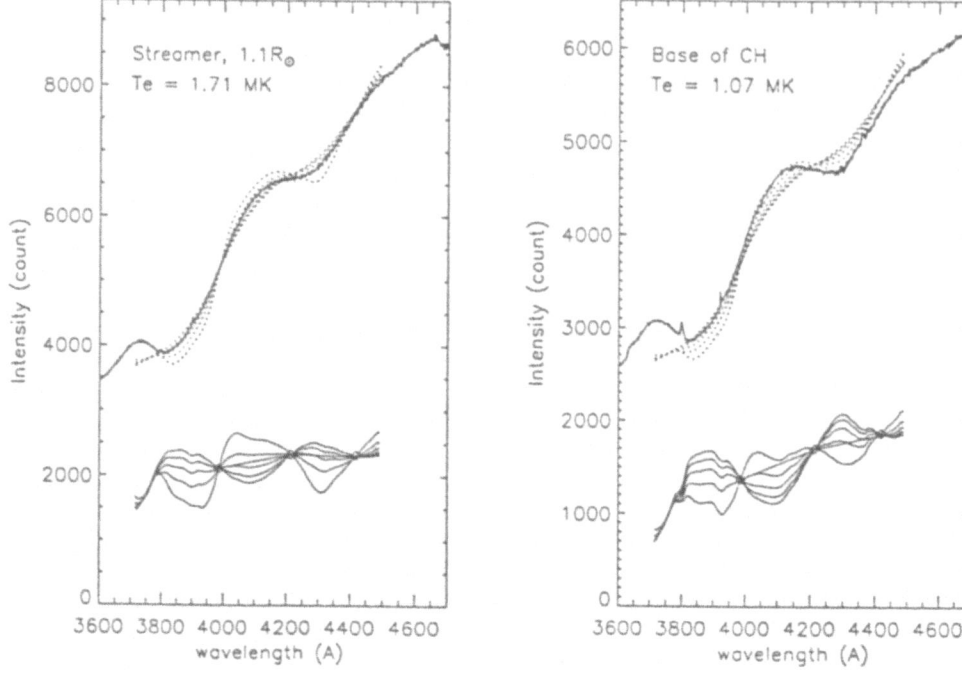

Figure 1. Examples of the continuous coronal spectra (solid curves) for the streamer (left) and the base of the coronal hole (right). Theoretical curves for T_e =0.6, 1.2, 1.8, 2.4, and 3.0 MK are indicated by dotted curves. The bottom curves show the difference between the observed and the calculated spectra indicating the goodness of the fitting with values multiplied by 2 and offset by a constant. Minimizing the difference gives an electron temperature of 1.71 and 1.07 MK for each spectra.

electron temperature. The electron temperature T_e was thus determined for each coronal point on the slit of the spectrograph. Figure 2 shows the temperature distribution along the slit for three different locations on the sky. The intensity distributions of the continuum and Ni XII 4323Å emission line are also shown by the thick and thin solid curves.

It is found that the derived electron temperature is dependent on the assumed geometry of the corona because the Doppler effect in the Thomson scattering changes with the angle of the scattering. Results for the non-spherical coronal model are also shown in Figure 2 by the small asteriscs; for which we assumed that the streamer extends 15° behind the east limb.

The characteristics are summarized as follows:

1) The electron temperature T_e of the eastern streamer was estimated to be about 1.7-2.0 MK using the spherical coronal model. If we take into account a non-spherical density distribution of the streamer, the estimated temperature is reduced to ~ 1.6 MK.

2) A height variation of the electron temperature in the streamer was not detected within the error of the measurement.

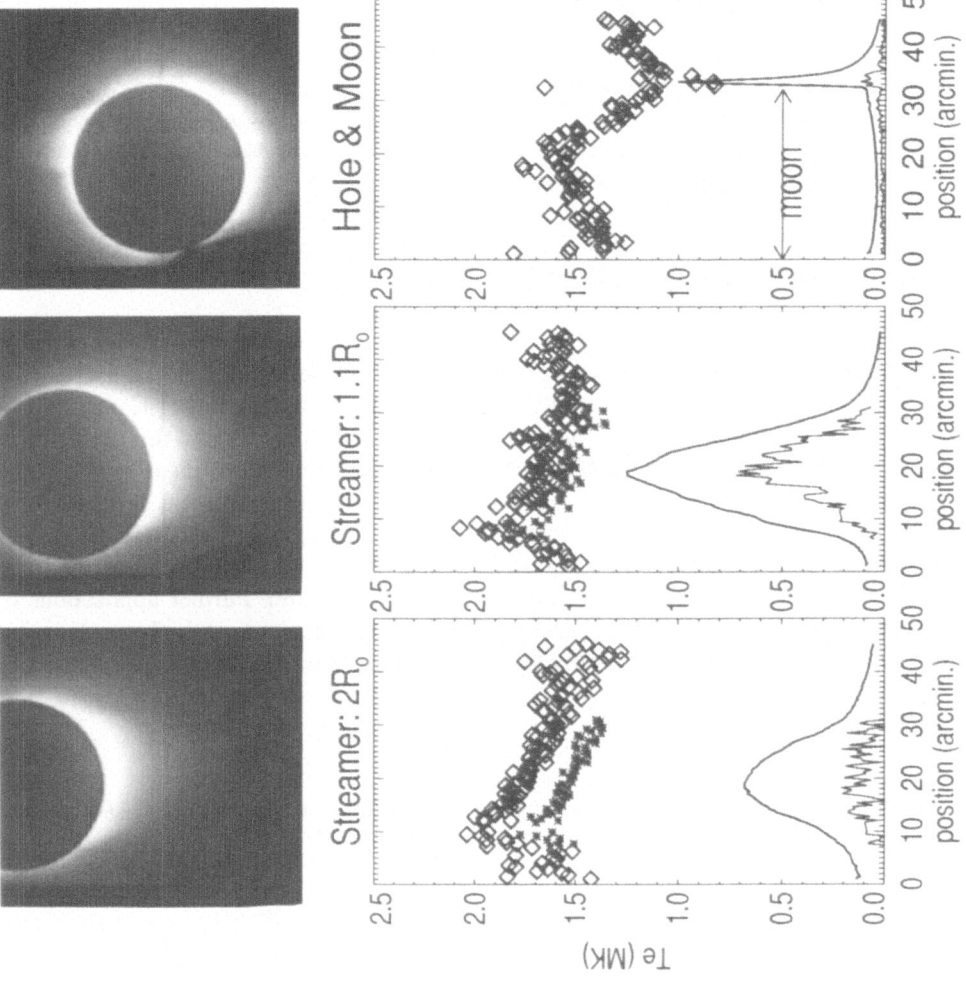

Figure 2. Temperature distributions along the slit at 3 different locations on the sky. The top shows the slit-jaw images. The open diamonds show the temperature derived from the spherical coronal model and the small asterisks show the temperature derived from the non-spherical coronal model for the streamer. The solid and dotted curves at the bottom show the continuum and Ni XII 4232Å mission-line intensities.

3) The base of the southern coronal hole obviously shows a lower temperature (0.9-1.1 MK) than the coronal streamer. T_e in the coronal hole shows a rapid increase with height, though this apparent increase can be partly attributed to scattered light from the sky.

We have also examined the distribution of the emission-line componentsand the entire wavelength shift of the continuous spectra in the streamer. The following results are derived from these analyses.

4) The emission-line intensities give fairly consistent results on the coronal temperature with those from the continuous spectra.

5) An evidence of the acceleration of the expanding motion between 1.1 and 2.0 R_O by ~ 80 km s^{-1} was found in the streamer.

The method for diagnosing the coronal electron temperature from the shape of the continuous spectra is superior to various techniques so far used for measuring the coronal electron temperature in the sense that: 1) fewer assumptions are required to determine the temperatures, the only critical requirement is to know the geometry of the coronal structure; 2) it enables one to study the temperature distribution in the extended corona ($R \geq 2\ R_\odot$) and; 3) only a quite simple instrument is required to achieve the measurement, though low-level of scattered light and accurate calibration is desired. The continuous coronal spectra also offer a chance for measuring the outflow velocity at the acceleration region of the solar wind. Further applications of this technique to study the coronal electron temperature and the solar-wind acceleration should be a challenging subject for future observations of solar eclipses.

References

1. Cram L.E., (1976), *Solar Phys.*, **48**, 3.
2. Ichimoto, K., Kumagai, K., Sano, I., Kobiki, T., Sakurai, T. and Munoz, A. (1996), *Publ.Astron.Soc.Japan*, **48**.545.

HIGH RESOLUTION CORONAL IMAGING
11 July, 1991 Total Eclipse with University of Hawaii
2.24 meter Telescope on Mauna Kea

B. LABONTE
University of Hawaii, Inst. for Astronomy
2680 Woodlawn Drive, Honolulu, HI 96822, USA

W. LIVINGSTON, J. ZIRKER
National Solar Observatory, NOAO PO Box 26732,
Tucson, AZ 85726; USA

Abstract We observed the corona with the potential for unprecedented spatial resolution afforded by a large stellar instrument at a prime site. An ambitious experiment involving the recording of 5 narrow-band 1024x1024 CCD images failed because of a last minute software change. Eclipses are unforgiving in this regard. However, a simple video camera was also fed by a beam-splitter pickoff with excellent results. This video tape, shown at the workshop,indicates how the field of interest was selected, concentrating on a large prominence at the NW limb. The role of seeing is evident; the smallest features detected were 0.6 arc-sec in size.

1. Introduction

The 11 July 1991 passage of the Moon's shadow directly across Mauna Kea and its cluster of large telescopes was a unique opportunity to search for coronal fine structure. Two major experiments were aimed at such an objective in visible light: Barry LaBonte led a team at the UH 2.24 m Telescope and Serge Koutchmy conducted similar experiments at the 3.6 m CFHT. An account of the latter is given by Koutchmy et al [1] and at this conference. Nothing has been published about the LaBonte experiments because of a near fatal data recording problem as explained below.

2. The Experiment as Planned

Briefly, CCD camera #1 had a field-of-view of 170 x 170 arc sec, 1/6 arc sec per pixel and a filter wheel providing the following:

Wavelength	Passband
5303Å	3Å
5500	200
5694	3
6394	3

Z. Mouradian and M. Stavinschi (eds.),
Theoretical and Observational Problems Related to Solar Eclipses, 35-37.
©1997 *Kluwer Academic Publishers.*

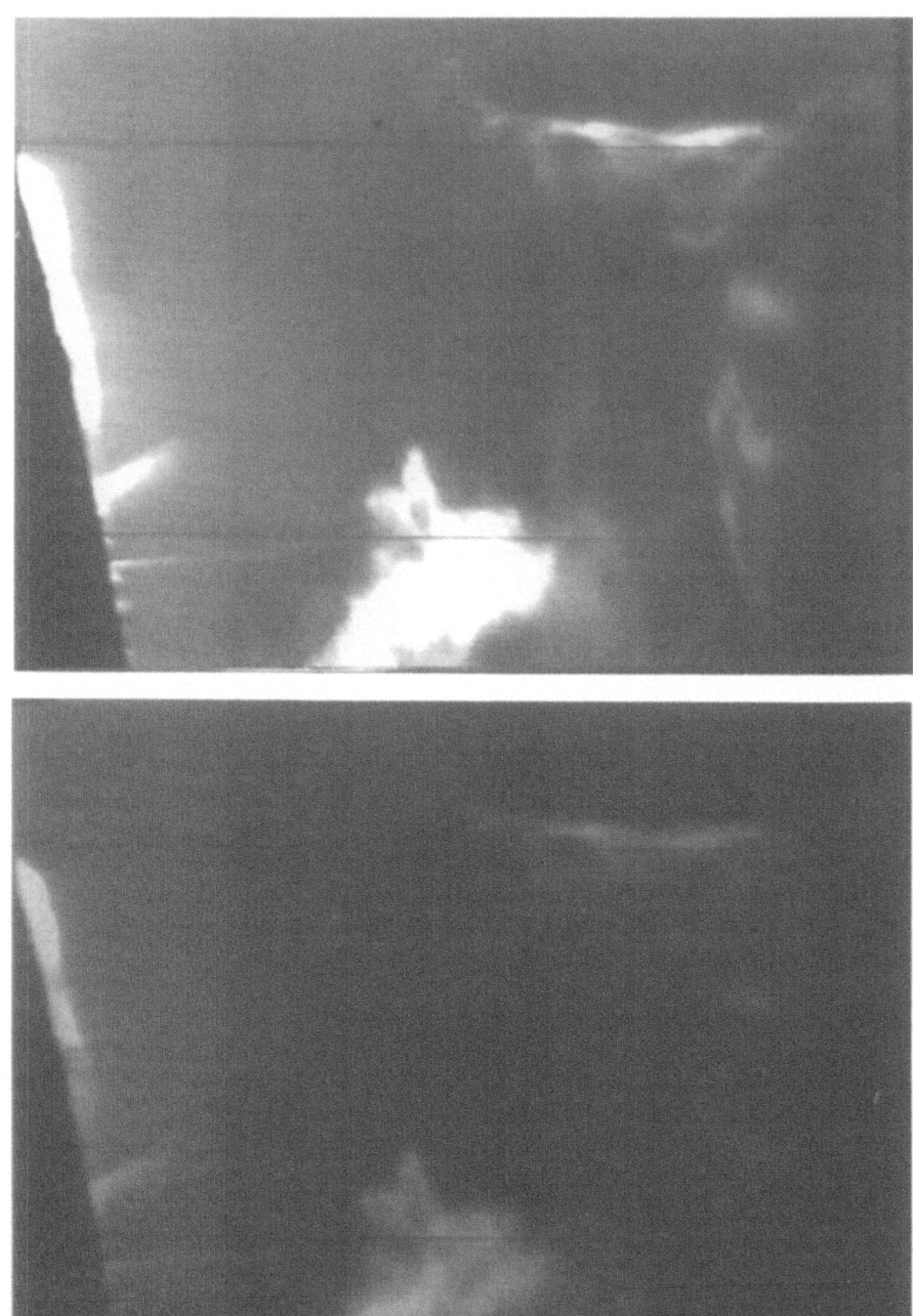

Figure 1. Video taken in the light of He 4771Å plus continuum. Upper display set to show fainter features; lower the brighter ones.

CCD camera #2 was devoted to Hβ (4861 Å) with a passband of 100 Å and a resolution of 0.3 arc-sec per pixel. A beam-splitter also provided video from a CCD Cohu camera at the same resolution looking at 4471 Å with a bandwidth of 98 Å. This means it was centered on the emission of He I 4471.5 Å, a triplet with the same lower level as He D3. There are 7000 frames.

Taking elaborate care to avoid pointing at the Sun during the partial phases, the experiment during actual eclipse came off without a major hitch. Data was recorded on computer disk. It was this last step which proved faulty; instead of reading CCD images 64 rows at a time and advancing through the process 16 times to cover 1024 pixels, the first 64 rows were read repeatedly 16 times! Unfortunately, an eclipse is a one time event.

3. Results

There may well be interesting information in those first 64 rows but, disappointed by the recording failure, the narrow band data have not yet been studied. Figure 1 shows a single video frame taken near the end of the eclipse just as third contact approached. We reproduce two versions of the same frame with the dark levels set to saturate in the first (better to see the fainter structure), and set higher in the second to reveal unsaturated detail in the brighter parts. Mainly this is light from helium emission plus some continuum. A narrow ray is seen near the bottom. Most of the scene is occupied by a large presumably active prominence. Watching the video one is aware of diffuse background structure which extends off out of the reproduced frame to the right. The eye can integrate and see such detail that is lost to a frame-grabber image. The smallest features detected were about 0.6 arc sec. This is consistent with night-time seeing on the previous nights.

Acknowledgments

During preparations and at the telescope we are grateful to Jim Anuskiewicz, Kimberly Leka, and Frank Cheigh. Mike Doe made the prints reproduced here. NSO is operated by the Association of Universities for Research in Astronomy, Inc., under cooperative agreement with the National Science Foundation.

References

1. Koutchmy, S., Belmahdi, M., Coulter, R.L., Demoulin, P., Gaizauskas, V., MacQueen, R.M., Monnet, Mouette, J., Noëns, J.C., November, L.J., Noyes, R.W., Sime, D.G., Sovka, J., Vial, J.C., Zimmermann, J.P., Zirker, J.B. (1994) *CFHT* eclipse observation of the very fine-scale solar corona, *Astron. Astrophys.*, **281**, 249-257.

CORONAL PHYSICS INFERRED FROM THE ANALYSIS OF FINE STRUCTURES

S. KOUTCHMY
Institut d'Astrophysique de Paris, CNRS,
98bis Bd Arago, 75014 Paris, France
associated with the Paris VI University

Abstract. Solar total eclipses are rare occasions offering the opportunity to make a snapshot of the solar corona. The availability of a large radiative flux makes sophisticated methods usefull to analyse all parts of the highly structured white-light corona. Coronal absolute intensities and line emissions, including their polarization, are also studied to analyse density and temperature inhomogeneities, velocities and magnetic fields. Detailed density distribution is directly extracted for fine coronal structures. During the July 11, 1991 eclipse, the large 3.6m aperture CFH optical telescope was used to analyse time sequences over small coronal fields and to image the finest structures never seen ; results from this experiment are presented with emphasis on small-scale dynamical plasma processes including wave-phenomena. Finally, we give an overview of what is known on coronal structures from eclipse observations : temperatures, densities, velocities, occurrence and structure of streamers, coronal holes, threads, etc.

1. Introduction

During total solar eclipses, when first the photosphere and then the chromosphere are completely hidden from view, a faint white polarized halo produced by the scattering of the solar light from electrons (due to the K-corona) becomes visible : the white-light (WL) corona. Around the moving disc of the Moon slightly larger than the Sun, the corona can be analyzed during a few minutes from the transition region (TR) up to the most extended structures. At fair spatial resolution "snapshot" observations are made, including spectra, to analyse the temperature and velocities. No shadowing or vignetting of the WL corona is seen during a natural eclipse and in good observing conditions, the limitations in the field of view is essentially due to the dominance of the brightness of the dust F-corona, [1,2,4,7,17].

Absolute photometric measurements can be performed with a high accuracy, thanks to images of calibration stars [1,2], simultaneously observed on the same picture as the corona. Suitably attenuated solar light can also be used immediately before or after the totality. Progress made in the last decade includes the use of an accurately calibrated radial neutral density filter to compensate the large average radial gradient [1,2] of the WL corona and to precisely remove parasitic effects produced in the instrument and the sky (the aureola effect) ; the use of such filters considerably improves the spatial resolution by avoiding overexposures. Strictly linear and highly sensitive detector systems are also used, especially for making time sequences with narrow passband filters. However, because of its simplicity, speed, reliability and large

Z. Mouradian and M. Stavinschi (eds.),
Theoretical and Observational Problems Related to Solar Eclipses, 39-52.
©1997 *Kluwer Academic Publishers.*

storage capability, the film is still largely in use at eclipses. We believe that to reach a higher spatial resolution during an eclipse, fast photographic imaging and video CCD imaging in WL (preferably in colors!) potentially represent very valuable methods. Amateur-class telescopes can easily achieve sub-arcsec resolution outside of eclipses (Lunar studies, for ex.) and there is no serious reasons to assume that this is not possible during the totality, provided a long enough focal length is used.

Regarding optical spectroscopy, many emission lines are easily observed with a slit-less spectrograph. However, it is today more useful to look with modern detectors at more "extended" diagnostics which are difficult to perform in space : full line profile analysis of faint parts, like coronal holes and streamers at r $\geq$2, study of Fraunhofer-lines scattered by the K and the neutral component. Because of the lack of photons, X-EUV spectroscopy can hardly provide information coming from the analysis of an emission line at r$\geq$1.5 ; Doppler shifts, line-width (to measure turbulent velocities) and even the radial gradients are there poorly known ; the Lyot-coronograph is also, for obvious reasons, improper for that, so total eclipses still offer the best opportunity for using the most advanced diagnostics for testing the theory of heating and acceleration of coronal plasma. Unfortunately, total eclipses are rare and rather short. Because the coronal activity changes from one eclipse to an another one, it is difficult to check the result of a "snapshot", so results of the most sophisticated experiments are sometimes disappointing and even, not reported ! Finally, no information is provided over the solar disc during an eclipse. Fortunately, soft X-Rays (Yohkoh) and EUV coronographs (EIT on SOHO) very successfuly today supply these observations.

There are well established and classical results from eclipses: WL-photometry and polarimetry, structural analysis. They cover many solar cycles [3], so these eclipse observations are invaluable for analysing the coronal activity solar cycle, although coronal mass ejections (CME) have rarely been observed at eclipses. Obviously, orbital externally-occulted WL-coronographs [18] are more suitable instruments to observe CME, so we will not extensively consider here these events. More results will becoming soon from the highly successful operations of two new WL coronographs (LASCO) flown on SOHO. Finally, let us notice that the best spatial resolution observations of the corona were performed with the 3.6m aperture Canada-France Hawaii Telescope (CFHT) on July 11, 1991, bringing a new insight in the study of the fine scale phenomena important for the physics of small-scale processes [21].

2. Model of Coronal Structures

2.1. MORPHOLOGICAL ANALYSIS

It is generally accepted that the analysis of quasi-stationary coronal structures of the Sun (including to some extent small-scale structures) gives an important insight into the knowledge of the solar magnetic fields, their evolution, their origin and, accordingly, of the solar activity. Indeed, it is quite obvious that a low beta plasma ($\beta=8\pi$ P/B^2) behaves like a tracer of the lines of force of the magnetic field when a dynamical corona is considered. However, it is not always the case, especially at fine scales. The large value of the electrical (and thermal) conductivity of the coronal plasma is an additional factor to be considered. It is apparent on eclipse pictures that structures overlying the coronal hole regions are open, as are the magnetic field lines of force ; conversely, arch-like and loop-like structures are well observed around active regions and filaments [4] corresponding to closed magnetic field lines which will eventually be «disrupted» during a CME.

White-light eclipse pictures do not properly record the extended coronal features unless special filters are used (radial neutral filter). An off-focus circular mask is also sometimes used [22]. Then, coronal details appear more clearly. However, the overlapping of the details superposed along the line of sight and, also, the large differences often observed between equatorial and polar regions, make the pictures still different from what will be "seen" by an experienced observer. This is illustrated on the Figure reproduced in [2,3,4] which shows, on the example of the well studied eclipse observation of June 30, 1973, a processed image which leads to a result not far from the interpretation given when a drawing is produced from a naked eye observation. Such a WL picture shows best the distribution of densities in the corona, because the modulations are linearly related to variations of the amount of free electrons and, consequently, to the plasma densities, integrated along the line of sight. It is quite evident, from the inspection of such figure, that the plasma is largely confined to small-scale structures, that is to thread-like filaments and loops. This impression is largely confirmed by the inspection of an good resolution X-ray picture which shows emissions of the inner corona observed on the solar disc.

Plasma densities are directly deduced from WL observations (see further) ; temperature structures can be analysed with filtergrams made in line emissions. Both the richness and the complexity of structures are evident, suggesting permanent dynamical phenomena (flows) occuring essentially at very small scales and sudden instabilities occuring at large scales. Although all scales are largely revealed in these numerous observations, an even better spatial resolution is required to fully understand the physics of these structures. Attempts to improve the resolution were made recently [21] by processing the WL pictures, see Figure 1 obtained with the CFHT. This is the best resolution image of the corona never obtained. Model calculations can help in analysing the morphology of the corona. Let us consider the pressure equilibrium of the coronal plasma. Briefly, assuming an isothermal corona, the gas pressure P in the structures can balance the magnetic field pressure, at least in the middle corona where $P \sim r^{-6}$, because the amplitude B of the large-scale magnitude field (low-order harmonic components of the photospheric field) decreases like r^{-3} there. Using a potential field approximation and some practical boundary layer conditions, numerical methods have been largely applied to "extrapolate" the magnetic field measured at photospheric levels. The deduced global morphology mimics the observed density structures [5]. At very large scales, the approximation is probably correct. However, the corona is filled with very small-scale structures like jets and threads, and the changing beta of the plasma is not known. Therefore, a rather detailed knowledge of the plasma density distribution is required to calculate the magnetic field needed to balance the lateral gas pressure. Loops are a special case as a dynamical equilibrium should be considered and the field is likely to be non-potential ; there, temperature inhomogeneities play a hidden role ; the so-called scaling law relating the temperatures, the pressure and the loop length, prevails. Tangential discontinuities (TD) are an other case.

At first sight, it is tempting to consider the temperature responsible for the variations of the heights of arches observed in the intermediate corona, outside active regions or close but not above active region. There are noticeable variations during the solar cycle. The radial variation of electron densities is presumably related to the temperature ; however, arches are clearly larger of size at sunspot minimum when the coronal temperature, presumably related to the strength of the solar magnetic field, is lower, so we suspect that the morphology of the corona is probably more dependent on the evolution of the large-scale solar magnetic field; it is not directly related to thermodynamical parameters like the temperature or, even, to the quasi-static magnetic field, but to their changes. It makes the physics of arches definitely different from the one of loops where the scaling law dominates.

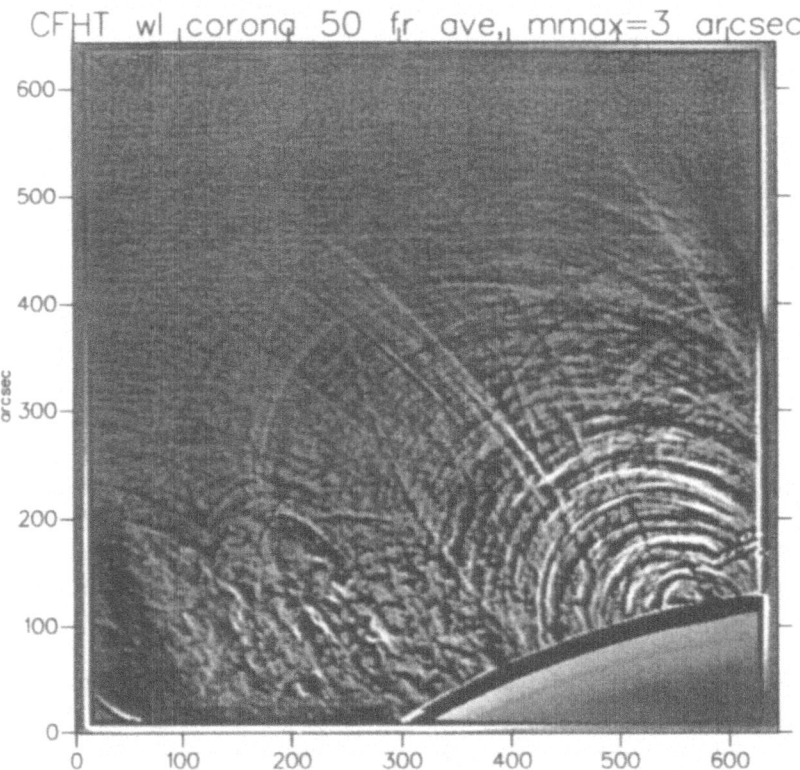

Figure 1. A detail of the inner corona structures from one of the CFHT exp-t, during the July 11, 1991 solar total eclipse. This is probably the best resolution WL coronal picture never obtained. It has been processed to show the main features using the Madmax algorithm [21]

Although the occurrence of many loops above and nearby active regions strongly suggests the coronal plasma frozen in magnetic lines anchored in the Sun, there are many locations in the intermediate and outer corona where densities are largely enhanced above neutral lines and sheets extended outside active regions; the most extreme cases are the large equatorial streamers at sunspot minimum. This equatorial "belt" of streamers is believed to be the source of the heliospheric sheet [5,6]; however, streamers are often seen at high latitudes and even at poles when the magnetic field is reversing its sign of dominant polarity. Accordingly, it is important to consider the morphology of the corona and the solar cycle [3]. A successful model of the heliosheet made of streamers is proposed in [22].

This is in contrast with the map deduced from the computation of "magnetic field lines" using the current free approximation (potential field) and the observed magnetic fields at the photospheric level; a detailed analysis of such maps should be made to match the observed individual structures of the intermediate corona. The occurrence of very large azimutal and/or tangential gradients of pressure (TD) far overpassing the radial gradients, is another difficulty when potential fields are considered [6,20].

2.2 THE RELATIONSHIP BETWEEN THE CORONAL STRUCTURES AND THE CORONAL MAGNETIC FIELD

To explain the large transverse gradient of gas pressure observed in the corona [10,12,16,17], we know only one force capable to balance the corresponding large pressure forces : the magnetic force $\vec{J} \wedge \vec{B}$, sometimes also called the Ampere force (when net currents are implied) or the Lorentz force (when a single charge is considered). To account for electric currents inside a plasma is difficult ; we first have the magnetic field induced in the corona by the most powerfull currents : those due to flows in the dense layers of the photosphere. Let us consider an example. To roughly match the field induced above a simple active region made of sunspots and faculae, we use an axi-symmetric model with a double horizontal current ring which permits to compute precisely the magnetic field (potential) outside (see Figure 2). The computed structure match quite closely the typical coronal structures but what is interesting is the occurrence of special surfaces, called separatrices, and especially, of a singular or neutral point. There, the magnetic field is vanishing ; accordingly, this region of the corona can be regarded as a trapping region for the coronal plasma. Under special circumstances, annihilation of magnetic energy can occure in this region because rapid changes of configuration and "reconnection" of lines, but this is a whole new field and is outside the scope of this paper.

2.3 DENSITY STRUCTURES DEDUCED FROM ECLIPSE PHOTOMETRY

Since the intensity of the WL corona is linearly related to the free electron densities due to the Thomson scattering, eclipse photometry provides direct measurements of coronal densities. However, a good absolute intensity calibration is required and the

44

contribution from the F-corona due to the scattering on dust particles should be subtracted. In first approximation, the coronal atmosphere is considered homogeneous and the measurement of the polarization ratio greatly help the separation of the K and of the F components, because the F-corona is not polarized for $r \leq 5$; the background A due to the parasitic light (instrument, sky) is not considered polarized [1,2] and the polarization ratio P_K of a model homogeneous K-corona can easily be predicted [7]

Although average density distributions are useful for comparing the corona at different latitudes and different phases of the sunspot cycle, they are far from the real values. The most conservative estimate of the filling factors (f.f.) in the intermediate corona gives a value of order of 10%, so average values correspond to a vastly oversimplified picture of the corona, even without considering really small-scale structures with diameter less than 10 arcsec (threads).

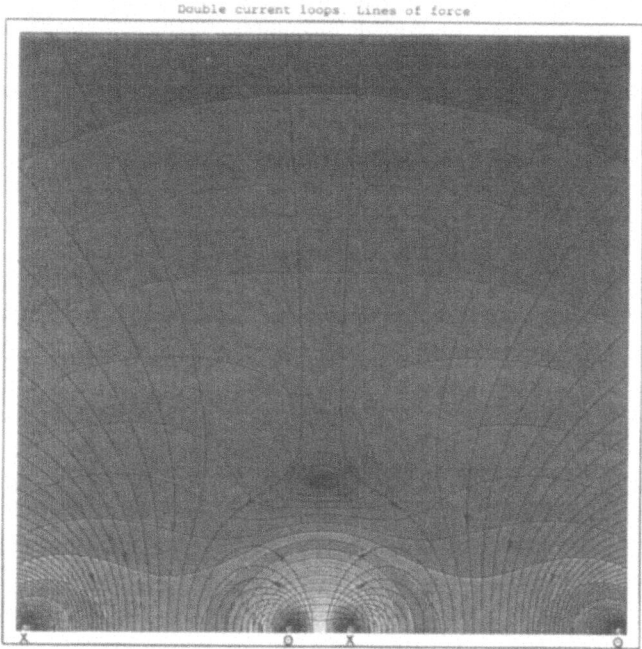

Figure .2 The set of magnetic field lines and lines of equal values of the strength of the magnetic field (superposed in gray scale) for an axially symmetric system of current rings. Notice the singular or neutral point on the axis, which is a trapping region for the coronal plasma, and the separatrices.

Let us consider the surface local brightness (intensity) of the K-corona $I=I(\rho)$ along a radial direction, ρ being the projected on the plane of the sky radial distances. It has been found convenient to represent $I(\rho)$ by a sum of terms of the form ρ^{-m}, with m taking different predominant values (from 17 to 3.5), depending on the value of the radial distance. Between the surface and few solar radii ($r < 3$ or 4), the derived radial density distribution, with $m_1=17$ and $m_2 = 7$, follows a curve described by a hydrostatic law with a constant temperature T_h of 1 to 2 x 10^6 K (10^6 K=1 MK), see Figure 2. This seems to indicate that the departure from hydrostatic equilibrium due to the hydrodynamic flow (solar wind) is small in this region. Because the lack of place in

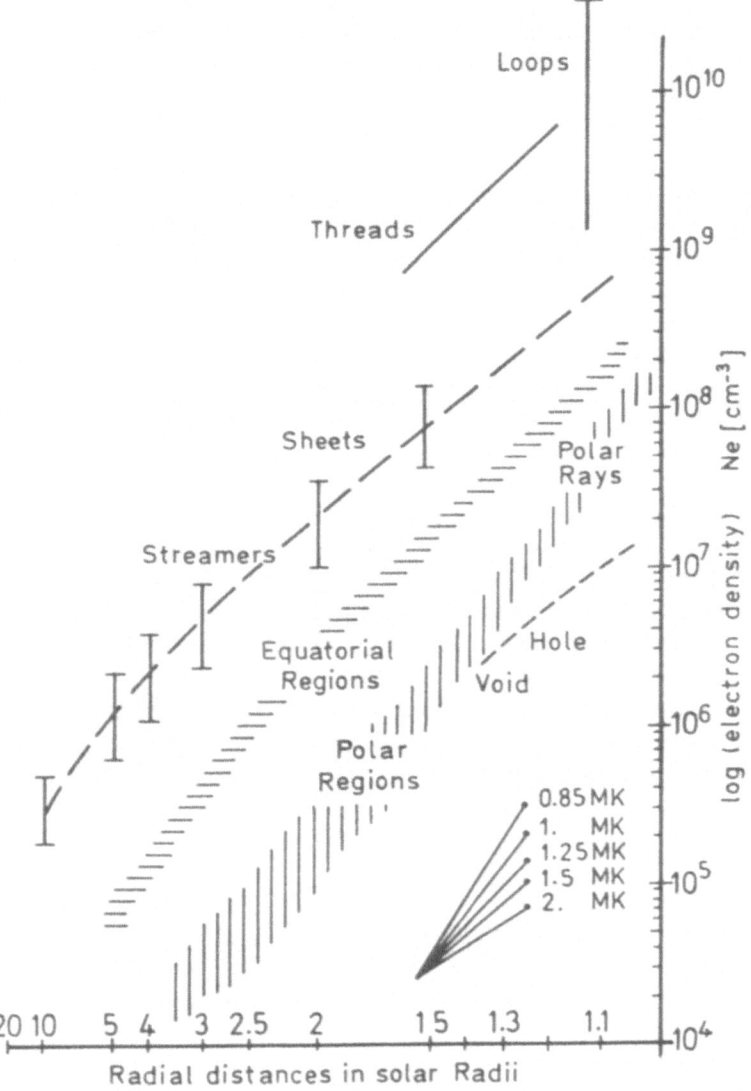

Figure 3. Radial variations of the electron density in different parts of the corona from eclipse photometry. Inserted at the bottom, note the graph with a set of lines to show the slopes corresponding to different hydrostatic temperatures. Densities which are shown are extreme values but for the Equatorial Regions, were average values are given. Threads are small cross-sections (<10") structures.

this paper, we refer to formula given on papers [7,17] and [21] to justify Figure 2. It is useful to define l_{eff} as the length of the l.o.s. which correspond to a decrease of the scattered intensity by a factor $e \simeq 2.7$ compared to the value obtained in the plane of the sky ; assuming some radial variation of $N_e(r)$ and taking an hydrostatic temperature $T=1.5$ MK, l_{eff} can be deduced, see [17,21]. Now, the filling factor f.f. in its simplest form could also be introduced. Assume that along l_{eff} only a part α of the l.o.s., is occupied by a plasma of constant density, then the f.f. is equal to α, assuming a homogeneous distribution normal to the l.o.s., which is not exactly the case because structures have a limited extend, so f.f.$\leq \alpha$. Another factor which is often mentioned in the literature to describe the density irregularity is : $< N^2_e > / < N_e >^2$. There is a relation between this last factor and the f.f. ; however, because a whole spectrum of irregularities exists in the corona and observations have a finite resolution, this relationship is not easy to deduce, nor to measure. Note that permitted line emissions (and radio emissions) are also a function of the coronal local temperature and not just of $< N^2_e >$; forbidden lines emission in the visible is also function of a power close to 2 of electron densities.

Even a coronal hole is very inhomogeneous, with plumes, threads and jets, as clearly appearing on a good eclipse processed picture [2,4]. The most popular model of a coronal hole based on the ATM-SkyLab data, disagrees with this eclipse measurements and does not take into account inhomogeneities ; indeed when structures are considered, the problem with C.H. is completely open and a lot of efforts will have to be made at eclipses to resolve the discrepancies. Hopefully, results which will be collected with the LASCO coronographs flying on board the SOHO spacecraft will also bring a new insight.

A last special case is streamers which are very easy to study at eclipses. The contribution of a streamer which is close to the plane of the limb can easily be measured as a large modulation ΔI of the azimutal distribution of intensities ; measuring the corresponding polarization degree, it is also possible to evaluate the geometric thickness Δl along the l.o.s. of the streamer when is not too large. Therefore, we have a more simple relationship:

$$\Delta I(\rho) = B_\odot \ \Sigma \ W_\lambda(r) \ N_e(r) \ \Delta l \qquad (1)$$

A full analysis of such case has been presented in /8/. There, the distribution of N_e in the cross-section of the streamer is given by :

$$N_e = cte \ exp \ [-(x^2 + \eta y^2) \ \sigma^{-2}] \ exp \ (-gz) \qquad (2)$$

where η describes the ratio between the extension along the l.o.s. and the thickness observed in the plane of the picture and σ is the equivalent cross-section. Because a large streamer is made of several overlapping parts (sheets), it is impossible to consider in detail the distribution of N_e; so a gaussian shaped cross-section with $\eta \cong 1$ is sometimes used as a first approximation. The deduced densities along the axis of the streamer are shown on Figure 2 ; a remarkable change of the behaviour of N_e is detected in the range $r=3$ to 3.5 which corresponds to the smallest values of the effective cross-section σ. Assuming this part corresponds to the crossing of the sound velocity by the outwardly expanding gas "inside" the streamer (by analogy with the Laval-nozzle) and using the equation of mass conservation, we deduce :

$$N_e(r)\ \sigma(r)\ \upsilon(r) = \text{cte} \qquad (3)$$

which permits to readily compute $\upsilon(r)$. An increasing velocity, up to the limit of visibility of the stalk of the streamer, is obtained ; therefore, no curvature is predicted nor observed with the present precision. This procedure was repeated more recently on the example of a narrower and longer streamer observed at the 1973 eclipse and the deduced velocities were given in [17]. Again, no curvature is observed. However, the deduced velocities are definitely decreasing after $r=5$ and taking into account the position angle of the streamer, a curvature would have been observed if the gas was expelled freely and the magnetic field frozen in. This effect is not yet observed...

3. Very Fine Structure

Although the inhomogeneity of the quiet solar corona has been recognized for a long time, no concrete account of it have been made in the theory of the acceleration of the solar wind nor in the theories of extended coronal heating. Theoretical works are essentially based on time and space averages. A good coronal eclipse picture shows that the coronal plasma is indeed confined in fine scale structures; eventually, these structures clump together locally to form a far larger scale structure like a coronal

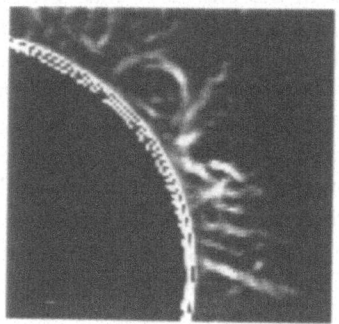 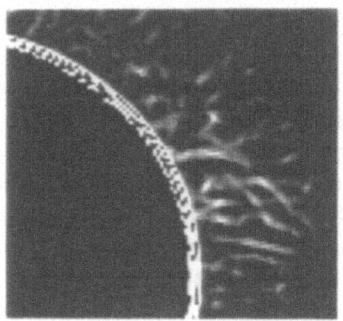

530.3 nm   637.4 nm

Figure4. Structure of the inner corona observed at different temperatures: at left, in the 530.3 nm line of Fe XIV formed near 2 MK and at right, in the 637.4 nm line of Fe X formed near 1 MK. Pictures were simultaneously obtained by J.Sykora at the 1981 eclipse and subsequently processed at NSO-SP to remove the background and improve the visibility of features (Madmax algorithm).

streamer. In that sense we say the corona is *striated* with fine scale elongated structures. The physics of the small-scale structures is obviously different: taking into account the filling factor, densities increase dramatically and their tangential gradients reach extremely high values suggesting discontinuities ; structures can be also detached or twisted. Finally, fine scale structures can interact, cross one each other, pointing to the reconnection - disconnection phenomena occuring at even smaller scales. The study of these small-scale structures is still rather frustrating ; there are several reasons for this, namely :

    a) The spectrum of fine scale inhomogeneities is unknown. Moreover, no statistical account can be furnished due to geometrical superposition or overlapping effects; structures are preferentially stretched radially, at least for $r \geq 1.5$.

    b) Up to now, only coronal loops and the rather extended streamers and coronal holes have received attention [6,9,21].

    c) Optical observations, which potentially could give the best spatial resolution, are mainly of "amateur-size" apertures, giving essentially unresolved observations. Moreover, the modern detectors like CCD cameras are not yet in use in coronal imagery. The 1991 total solar eclipse has been an exception because of the unique possibility to use the best optical telescopes in the world and modern detector-systems at Mauna Kea (Hawaii).

Before going further in studying a few cases of density inhomogeneities, let us consider temperature variations in the corona. It is well established that essentially 2 main classes of emission lines exist (we do not consider the class III lines formed at $T > 3$ MK which emit in very active region and micro-flares): class I corresponds to $T=1$ to 1.5 MK (typical lines are those of Fe IX,X,XI) and class II, corresponding to $T=1.5$ to 2.5 MK (Fe XIV,Fe XIII, Fe XV). On Figure 4, we display the emission observed at the 1981 eclipse in the green Fe XIV line and in the red Fe X line. Pictures were specially processed to increase the visibility of structures (octo-directional maximum of secondary derivatives): they clearly demonstrate the absence of obvious correlation between these temperature sensitive emissions, demonstrating the inhomogeneities in temperature. Examining the simultaneously observed picture in WL, which gives the distribution of N_e, we however noticed a correlation with the Fe XIV picture ($T \cong 2$ MK) which seems to be the most probable temperature in the corona outside active centers. The local pressure is not greatly affected by these changes of temperature, because density inhomogeneities are far more important. It is probably not the case in coronal flares observed in SXR which are not studied in this paper.

TABLE 1. Tangential Discontinuities: Typical Values using a 5 arcsec Resolution Picture

$r[R_\odot]$	1.50	2.0	2.40
$N[e^- / cm^3]$	1.1×10^8	1.7×10^7	0.4×10^7
$\delta l[Mm]$	4.0	4.5	5.0
$\nabla_\perp n[e^- /cm^4]$	0.25	0.027	0.006
$\nabla_r n[e^- / cm^4]$	0.77×10^{-3}	10^{-4}	$4. \times 10^{-5}$
ratio $\nabla_\perp/\nabla_r$	> 325	> 270	> 150
$\Delta B_{//}[gauss]$	0.8	0.25	0.12
$T_{hyd}[10^6 K]$	1.25	1.0	0.8
$R_L = \dfrac{m\bar{v}c}{eB_{//}} \left[cm\right]$	14	40	80

3.1. CORONAL SHEETS AND TANGENTIAL DISCONTINUITIES

When a good resolution coronal WL eclipse picture is examined, several slightly curved quasi-radial coronal sheets can be identified. Often, they are parts of a large coronal streamer. Almost inevitably they show, from one side, the very characteristic signature of a coronal discontinuity: a large and definite jump of N_e [10,20] ; Koutchmy identified these jumps of N_e with a tangential discontinuity (TD), as defined in plasma physics. Let us summarize the characteristics of a coronal TD, as obtained from the analysis of the characteristic case of a coronal sheet of known radius of curvature (note that we do not consider a "flat" sheet) : δl is the full observed «width» of the discontinuity ; it is close to the resolution of the picture ; so the jump of densities at the boundary could be higher. However, the deduced numbers are already very significant : the gradient of densities $\nabla_\perp$ across the TD is several orders of magnitude larger than the radial gradients ! Often an emptiness or "void" seems to exist along . The strength of the magnetic field was calculated from the conditions needed to have a stable TD; the stability has been confirmed at least in the intermediate corona, by comparing eclipse pictures obtained at several hour intervals. The problem of interacting TD has been considered in details in a theoretical paper [20].

Let us assume the temperature is constant across TD ; we note that the gyroradius of e^- : $R_L \ll \delta l$ and that the evolution of TD is rather slow. If N_1 and N_2 are densities ($N_1 \gg N_2$) at each side, υ is the velocity and B the magnetic field, we have :

$$v_\perp \equiv 0 \quad ; \quad v_{//1} \neq v_{//2}$$
$$B_\perp \equiv 0 \quad ; \quad B_{//1} \neq B_{//2} \; ; N_1 \neq N_2$$

The current $j_\perp$ across D is also zero because we have no flux of particles across ($N_1 v_{\perp 1}$ $= N_2 v_{\perp 2} = 0$), but a bulk displacement of the whole T.D. is not excluded. P_1 and P_2 being the pressure at each side we finally have :

$$P_1 + \frac{B_{//}^2}{8\pi} = P_2 + \frac{B_{//}^2}{8\pi}$$

Taking $\Delta B \cong B$, $P_j = 2N_{ej} k T_{hyd}$, we get the value $B_{//}$ given in the above table which is probably the best estimation of the true magnetic field in the intermediate corona. A more sophisticated analysis would include the equation $j = 1/4\pi \nabla \times B$, and processes in the interface if the current layer is significant.

3.2. RAYS, THREADS AND PLASMOIDS

These structures of the inner and of the intermediate corona are exclusively observed on very high resolution eclipse pictures. They seem to be an obvious coronal counterpart of localized transition region phenomena occuring everywhere : above miniature ephemeral active regions that presumably overlay enhanced chromospheric network regions (although inner network phenomena cannot be completely excluded), or above filaments, or in and especially around sunspots flaring active regions or finally above polar regions (fine structure of polar plumes and rays). There are early reports of visual and photographic detection of the inner coronal fine jets by japanese observers and, especially, the Kiev Observatoiry group ; the first photometric works were done in 1968-73 [11] and since that time, several new measurements were performed [12] confirming the early theoretical predictions [13]. From the analysis of several excellent plates obtained at past eclipses we got a set of measurements showing the transverse photometric "size" (or FWHM) is comparable to the extension along the l.o.s. ; then using formula (1), N_e can be obtained. From the estimation of the cross-section σ_o , with $\Delta_l \approx$ FWHM and using equation (1) (3), we get the variation of velocities V_o ; however more relaible values are deduced from the observed proper motion. This is rather seldom, because proper motions are difficult to measure and emission lines line-profile analysis gives result which are difficult to interpret.

The 1991 eclipse experiments at CFHT brought completely new results. Surprisingly, small-scale structures at sub-arcsec scales showed up as a highly variable phenomenon. Elongated coronal threads of < 1 Mm diameter are well detected above equatorial active regions at more than 100 Mm from the surface, with a life-time of order of 100 sec ; rays with a larger cross-section have larger life-time. Proper motions are obvious on threads but they seem to fade out and the background is slightly changing. A larger event than threads was also observed at CFHT, namely an *isolated* cloud of coronal plasma or plasmoid fading out in the ambiant corona with dynamical associated phenomena [14,19]. Precise measurements of the position of the center of gravity can be performed over the time sequence of more than 6000 video-CCD frames. Only the component of the proper motion projected on the plane of the limb is seen. Additionally, the cloud which is of 3 arcsec size at the beginning of the video time sequence is splitting-up several times and ephemeral thread-like structures are produced. Ultimately the cloud disappeared (at the end of the sequence), after 200 sec of time. Values of the measured proper motions υ_{tot} in the plane of the sky of order of 60 Km s^{-1} were reported [19]. This observation of a coronal plasmoid and dynamical associated phenomena needs a more thorough analysis but indicates that an important part of the corona (if not all the corona !) is in a dynamical state at sub-arcsec resolution. Ejected plasma clouds are initially diamagnetic [15] and because we see them fading out in the bachground corona, MHD small-scale phenomena should be at work and are probably responsible for the heating.

4. Conclusions

Future eclipse observations could still be an excellent complement to the ongoing missions YOHKOH and to coronal observations on SOHO, and also to prepare the Solar Probe Mission.

Because of the good to excellent photometric precision performed at total eclipses, an absolute calibration can be provided during a short interval. Good spatial resolution observations at large scale need the use of a radial neutral filter. However, future eclipse observations should now focus on experiments which are difficult to perform on a spaceborne platform: *line profile analysis* of line emissions in the *intermediate corona*, WL-corona analysis of the inner and intermediate corona at *high spatial resolution*, etc... Eclipse observations can yield more on the analysis of *very small-scale phenomena* like the plasmoïd phenomena observed in 1991 ; not only a good aperture telescope is needed, but also a sub-arcsec seeing should prevail ! Future spaceborne observations free of seeing effects could be more effective when considering that problem. A large aperture groundbased mirror coronograph [16] would however permit these observations ; IR adaptative optics system should be a part of that instrument which is now proposed as a future NSO/NOAO facility in USA (the "CLEAR" project).

Acknowledgments

Results presented here mainly reflect the collaborations that I benefited from many scientists whose names only partly appear in the references and to whom I am indebted. The typing was diligently performed by Marie-Claude Billiault.

REFERENCES

1. Lebecq, C.., Koutchmy, S. and Stellmacher, G. (1985) The 1981 Solar Total Eclipse: II. Global Absolute Photometrical Analysis, *Astron. Astrophys.* **152**, 157.
2. Koutchmy, S. (1975) Study the June 30, 1973 Trans-polar Coronal Hole, *Solar Physics* **51**, 399.
3. Loucif, M.L. and Koutchmy, S. (1989) Solar Cycle Variations of Coronal Structures, *Astron. Astrophys. Suppl.* **77** , 45.
4. Koutchmy, S. (1977) Solar Corona, in : *Illustrated Glossary for Solar and Solar-Terrestrial Physics,* (eds) A.Bruzek and C.J. Durrant, Reidel, 39.
5. Hoeksema, J.T. (1989) Extending the Sun's magnetic Field through the Three-dimensional Heliosphere, *Adv. Space Res..* **9 - 4**, 141.
6. Koutchmy, S. and. Livshits, M (1992) Coronal Streamers, *Space Science Rev.* **61**, 393.
7. Saito,K. (1972) A non-spherical Axisymmetric Model of the Solar K-corona of the Minimum Type, *Annals of the Tokyo Astron. Obs.,* ?XII, **2**, 53.
8. Koutchmy, S. (1972) Etude Hydrodynamique du Grand Jet Coronal observé à l'Eclipse du 7 mars 1970, *Solar Physics* **24**, 374.
9. Zirker, J. (1977) Coronal Holes and High Speed Wind Streams, Colorado Assoc. University Press.
10. Koutchmy, S. (1971) Un Modèle de Grand Jet Coronal avec Renforcement de Région Active, *Astron. Astrophys.* **13**, 79.
11. Koutchmy, S.and. Stellmacher, G. (1976) Photometric Study of Chromospheric and Coronal Spikes Observed during the Total Eclipse of 30 June, 1973, *Solar Physics* **49**, 253.
12. Koutchmy,S. (1988) Small-scale Coronal Structures, in : *Solar and Stellar Coronal Structures and Dynamics*, Proceedings of the NSO/SP 9th Workshop, R.C. Altrock (ed.), 208.
13. Alfven, H. (1963) On the Filamentary Structure of the Solar Corona, in : *The Solar Corona*, IAU Symp. 16, J.Evans (ed.), Acad. Press, 35.
14. Vial J.C., Koutchmy S. and the CFHT Team (1992) Evidence of Plasmoïd Ejection in the Corona from 1991 Eclipse Observations with the CFHT, in : *Proceedings of an ESA Workshop on Solar Physics and Astrophysics at Interferometry Resolution*, 87.
15. Pneuman, G.W. (1983) Ejection of Magnetic Fields from the Sun: Acceleration of a Solar Wind containing Diamagnetic Plasmoïds, *Astrophys. J.* **265**, 468.
16. Koutchmy, S. and Smartt R., (1989) High Resolution Observations of the Solar Corona: Why and How?, in : *Proceedings 10th NSO-SP Workshop* , O. v.d. Luhe (ed.), 560.
17. Koutchmy, S. (1992) Streamer Eclipse Observations, in : *1st SOHO Workshop*, Annapolis, Aug. 25-28 1992, ESA-SP.
18. Koutchmy, S. (1988) Spaceborne Coronography, *Space Science Rev.* **47**, 95.

19. Koutchmy, S. and the CFHT Team, (1994) CFHT Eclipse Observations of the very fine-scale Solar Corona, *Astron. Astrophys.* **281**, 249.

20. Grib,S.A., Koutchmy S. and. Sazanova V.N. (1996) Some MHD interactions in coronal structures, *Solar Physics*, (in press).

21. November, L.J. and Koutchmy, S., (1996) Coronal W-L dark loops and density fine structure, *Astrophys. J.*, **466**, 512.

22. Molodensky, M.M., Starkova, L. I., Koutchmy, S., Ershov, A.V.., (1996) 3-D Structure of Heliospheric Current Sheet Deduced from Eclipse Observations, in: ASP Conf. Series, **95**, 385.

RADIATIVE TRANSFER IN ATMOSPHERES WITH RANDOMLY DISTRIBUTED INHOMOGENEITIES

A.G. NIKOGHOSSIAN
Byurakan Astrophysical Observatory,
378433 Armenia

S. POJOGA
Astronomical Institute of Rumanian Academy of Sciences,
Str. Cutitul de Argint 5, 75212 Bucharest, Romania

Abstract. We consider the problem of determining the statistical properties of radiation that emerges from a medium with randomly distributed inhomogeneities when the scattering process is allowed. The analytical formalism we put forward is based on Ambartsumian's method of addition of layers. It is shown that certain specific features of the extreme-ultraviolet (EUV) spectra of quiescent prominences may be explained generally by the results obtained.

1. Introduction

It is currently clear that the consistent interpretation of emission by a variety of inhomogeneous structures on the Sun, such as chromospheric spicules, fibrils, coronal plumes and prominences, needs in a suitable theory for the transfer of radiation through a gas containing randomly distributed inhomogeneities. One procedure for calculating the statistical properties of the radiation emerging from a multicomponent LTE-atmosphere was presented in [1,2].

The present paper treats a more complex situation of the non-LTE atmosphere. For simplicity, we limit ourselves to the case of the conservative scattering that is of primary concern in analysing the spatial and time variations of the L_α line intensity for solar inhomogeneous patterns. In general, the implication of our study is the theoretical interpretation of the EUV spectra of quiescent prominences. In this connection, we shall also discuss the optically thin lines such as $\lambda\lambda$ 1336A CII, 977A CIII, 1032A OVI for which both the results of [2] and of our own approach, based on Ambartsumian's method of addition of layers, may be applied.

2. Statement of the Problem

For expository reasons, we focus our attention on the one-dimensional model of atmosphere adopted in [2]. Let we have a plane-parallel multicomponent medium containing initial energy sources. We suppose that the optical thickness of an individual layer τ and the power of energy sources B are random variables that take

Z. Mouradian and M. Stavinschi (eds.),
Theoretical and Observational Problems Related to Solar Eclipses, 53-57.
©1997 *Kluwer Academic Publishers.*

only one of two possible values: τ_1, τ_2 and, correspondingly, B_1, B_2. The probabilities associated with these two events are p_1 and p_2 $(= 1 - p_1)$. No radiation incident on the medium is assumed, though this can easily be included if necessary. We are interested in the statistical properties of outgoing radiation intensity, such as the probability distribution of intensity, the mean expectation and the relative mean square deviation (R.M.S.D.) depending on the number of layers N.

By making allowance for the scattering process, we greatly complicate the transfer problem inasmuch as the state of a gas at a given point is no longer specified completely by the local values of thermodynamic parameters as in the presence of local thermodynamic equilibrium (LTE), and depends on the radiation field throughout the atmosphere as well. This fact particularly implies that, in contrast to the LTE case, the mean intensity emerging from any part of multicomponent will be altered as a result of of a new layer added to it.

As a rule, the time-scale of quiescent prominence variations is much greater than that characterizing the scattering process of L_α-quanta. Therefore, when observing a given distribution of intensity over the set of pixels, we are essentially concerned with a set of different realisations of inhomogeneities corresponding to a definite deterministic problem.

3. The Method of Addition of Layers

The solution of the stochastic problem at hand assumes the application of a layer-adding procedure which, in its general non-LTE formulation, was developed by Ambartsumian [3] (see also [4]). Applied to the medium of two layers with the reflectance $\rho(\tau_i)$ and transmittance $q(\tau_i)$ $(i=1,2)$, for the intensities T_1 and T_2 of radiation outgoing from each boundary of the composite medium, the method yields

$$T_1 = f_1[1 + a(\tau_1, \tau_2)q(\tau_1)\rho(\tau_2)] + f_2 a(\tau_1, \tau_2)q(\tau_1) \qquad (1)$$
$$T_2 = f_1 a(\tau_1, \tau_2)q(\tau_2) + f_2[1 + a(\tau_1, \tau_2)q(\tau_2)\rho(\tau_1)]$$

where $a(\tau_1, \tau_2)=[1 - \rho(\tau_1)\rho(\tau_2)]^{-1}$; f_i are the fluxes in each direction produced by the layer containing energy sources of power B_i.

From equations (1) one can infer the basic features of the model multicomponent problem under consideration. The complexity of these equations is due to their specific nonlinear form with respect to the reflection and transmission coefficients. In practice, this makes it impossible the multiple application of them in attempting to derive a closed-form analytical solution. Note also that the fact of distinguishing between values of emerging intensities indicates the importance of the layers' order of arrangement in the medium.

4. Some Analytical Results

In this section, we shall limit ourselves to considering some results needed for the further discussion.

First, we consider the special case of pure scattering that is of importance in interpreting the formation of strong resonant lines, as the L_α line. The essential simplification of the problem is due to the fact that the overall radiative energy released within the atmosphere will escape it. Now we have $p(\tau_i) + q(\tau_i) = 1$; $p(\tau_i) = \tau_i/(\tau_i + 2)$, $q(\tau_i) = 2/(\tau_i + 2)$; $f_i = B_i\tau_i$ and equations (1) take a form

$$T_1 = f_1 + f_2 + \Delta; \quad T_2 = f_1 + f_2 - \Delta \qquad (2)$$

where $\Delta = (B_1 - B_2)[\tau_1\tau_2/(\tau_1 + \tau_2 + 2)]$.

It is physically obvious that the mean value of intensity $<I_N>$ emerging from an N-component atmosphere, obtained by averaging over the all possibilities is

$$<I_N> = N<I_1> \qquad (3)$$

where $<I_1> = p_1f_1 + p_2f_2$. The rigorous mathematical proof of this assertion may be carried out inductively. It is clearly discernible from equations (2) that the above non-commutativity of layers with respect to the observed intensities, is due to the difference in the power of the energy sources. This fact can be readily seen also from the closed-form solution for the R.M.S.D., denoted by

$$\delta_N = [p_1p_2/N<f>^2][(f_1 - f_2)^2 + \omega(\tau_1, \tau_2)(B_1 - B_2)^2] \qquad (4)$$

where $<f> = <B\tau> = p_1B_1\tau_1 + p_2B_2\tau_2$. The mathematical difficulties in deriving δ_N arise in finding an explicit expression for $\omega(\tau_1, \tau_2)$, though, for certain low values of N, this can be found directly. The right-hand side of the equation (4) reveals the difference in the physical nature of two items involved. While the first term is controlled by fluctuations in the number of layers realised with various physical properties, the second stems from variations in the arrangement of layers for a fixed proportion of the number of layers. It is reasonable to expect that the values of the product $B\tau$ are subject to random variations over a greater extent than the mere internal sources, so that the first term may be assumed to be no less compared to the second. In the limiting case of $B_1 \sim B_2$, the dispersion of the expected values of intensity results from variations in optical thickness of an atmosphere and is given by

$$\delta_N = (r/N)[(1 - \gamma)/(1 + r\gamma)]^2 \qquad (5)$$

where $\gamma = B_2\tau_2/B_1\tau_1$; $r = p_2/p_1$. Now we are led to the binomial distribution for the probability of observing one or another value of intensity: indeed the probability that the atmosphere will be composed of k layers with properties (B_1, τ_1) and $N-k$ layers of the second kind, hence emitting $I = kB_1\tau_1 + (N-k)B_2\tau_2$, is $C_N^k\, p_1^k\, p_2^{N-k}$.

In the case of LTE-atmosphere, $p(\tau_i) = 0$; $a(\tau_1, \tau_2) = 1$; $q(\tau_i) = exp(-\tau_i)$ and $f_i = B_i[1 - exp(-\tau_i)]$. Applying the layer-adding procedure and considering the fact that reflection from the component layers is absent, we are led to the results of [2]. Particularly, the optically thin LTE-atmosphere of N layers obeys the binomial distribution for the

56

intensity of observed radiation with the mean expectation and R.M.S.D. given by equations (3) and (5).

5. Discussion

The above models are undoubtedly too idealised to be applied to any realistic situation. Even so, the theoretical distributions derived for the emergent intensity may be regarded as necessary background for understanding some specific features of observational data, all the more so as the qualitative pattern predicted by the solutions obtained, remains to be valid for the more general formulation of the problem.

From our direct experience with the EUV images of quiescent prominences, we infer that the spatial distribution of intensity for the L_α line shows a symmetrical Gaussian-like profile (Fig.1) with relatively small dispersion ($\delta \sim 0.04 \div 0.07$), while the optically thin lines exhibit an asymmetrical distribution with high dispersion ($\delta \sim 0.1 \div 0.3$) and with a tail in the range of greater intensities. The latter is strongly reminiscent of the Poisson distribution. Gaussian and Poisson distributions for observed values of δ fit the observational profiles with rather high degree of confidence.

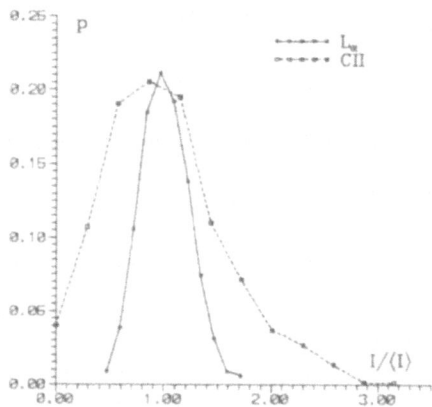

Figure 1. The typical intensity distributions for the L_α and $\lambda 1336$A CII lines.

Figure 2 displays the theoretical distributions for $\gamma = 0.1$ and $N = 12$. When $p_1 \approx p_2$, the binomial distribution differs only slightly from Gaussian even for relatively small values of N. On the other hand, when the atmosphere contains a small portion of strongly radiating elements (p_1 is small), we are led to a Poisson-like asymmetrical distribution in agreement with the observational profile. It is noteworthy that this kind of distribution may be obtained by assuming that the total number of structural elements varies randomly according to Poisson's law. This latter possibility leads, however, to somewhat low values of N as compared with the existing estimates and, on the contrary, yields greater values of N, being applied to optically thick lines.

Actually, both the above-mentioned reasons exist in combination, and each of them may become essential for the intensity distribution.

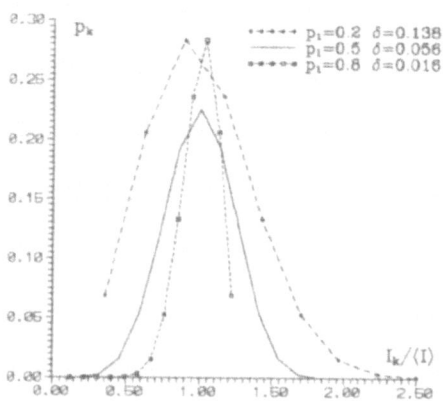

Figure 2. Theoretical distributions of the line radiation for $N = 12$, $\gamma = 0.1$
and indicated values of p_1 with corresponding values of δ.

Important information is contained in the observational values of δ. In light of high opacity of prominences in the L_α line, one may expect even lower values of than are usually observed. Now the formula (5) shows that δ ranges widely depending on γ (i.e. on the scale of inhomogeneities). Knowing γ we can infer of the dispersion in the amount of energy emitted by a gas, or ultimately in the emission measure. For small values of N, the symmetrical distribution for the L_α line implies that the structural elements with various physical properties are distributed uniformly. The asymmetry in the case of optically thin lines makes it possible to estimate the scale and the probability distribution of inhomogeneities.
The more detailed comparison of theoretical results with observational data is the subject of the separate work to be carried out.

Acknowledgements: A.G.N. acknowledges support of this research by NATO under grant CRG 960346 and S.P. was supported by the French Ministry of the Foreign Affairs.

References

1. Lindsey, C. (1987) *Ap.J.*, **320**, 893.
2. Jefferies, J.T.,Lindsey, C. (1988) Radiative Transfer in Inhomogeneous Atmospheres: a Statistical Approach, *Ap.J.*, **335**, 372.
3. Ambartsumian, V.A. (1960) *Scientific Works*, vol.1, Izd.Acad.Nauk Arm.SSR, Yerevan (in Russian).
4. Sobolev, V.V. (1963) *A Treatise on Radiative Transfer*, Princeton, van Nostrand.

PROBING THE SOLAR ATMOSPHERE THROUGH RADIOPHYSICS
A Review of Eclipse Radio Observations

M. MESSEROTTI
Trieste Astronomical Observatory
Via G.B. Tiepolo 11, I-34131, Trieste, Italy

1. Introduction

The solar corona is a highly ionized plasma which exhibits a complex variety of features on different space and time scales [1,2]. Such features range from sub-arcsec to solar radius scales and show lifetimes of the order seconds to days, indicating that the coronal plasma is basically non-homogeneous and subject to a set of physical processes, which, on one hand, determine such inhomogeneities and, on the other hand, are affected by them in a mutual feedback scheme. The recent progress in instrumental techniques from ground [3] and space [4, 5, 6] and the availability of multi-wavelength observations have been adding new tiles to this frame, but also rendered more puzzling the theoretical modelling on the different scales.

Many diagnostics [7] have been set up to derive coronal densities [8], temperatures [9], and magnetic field intensity [10] and topology [11], based on observations carried out in different spectral bands. In spite of that, many open problems still exist, to give just an example, about sub-arcsec fine structures, on the smallest scale, but even on the fine structure of large-scale features, on the largest scale [1,2].

Observations in the radio band can play a fundamental role in this frame. In fact, even if a completely self-consistent model does not exist yet for most radio phenomena [12], the kinetic theory and the radiophysics of waves' propagation allow to infer information on plasma density, polarity, intensity and topology of magnetic field [11].

In the following we will mainly consider radio observations performed during solar eclipses and emphasize the peculiarities of occultation techniques in the radio domain.

2. Tomography of plasma layers through solar radio observations

The relation which links the local electron plasma frequency f_{pe} with the local electron density N_e is

Z. Mouradian and M. Stavinschi (eds.),
Theoretical and Observational Problems Related to Solar Eclipses, 59-67.
©1997 *Kluwer Academic Publishers.*

$$f_{pe}(r,\vartheta,\varphi) = f_{pe}\big(N_e(r,\vartheta,\varphi)\big) = 8973 \cdot 10^{-6} \sqrt{N_e(r,\vartheta,\varphi)} \qquad (1)$$

with $f_{pe}(r,\vartheta,\varphi)$ in MHz, $N_e(r,\vartheta,\varphi)$ in cm^{-3}, in a Sun-centered, spherical reference frame. A radio wave with frequency f can propagate only if $f > f_{pe}$, as at $f = f_{pe}$ the refraction index vanishes causing reflection and at $f < f_{pe}$ it becomes imaginary causing absorption. The electron density is decreasing with increasing radial distance r and hence the critical frequency f_{pe}: multi-frequency radio observations map radio emissions occurring in different plasma layers in the solar atmosphere, located at increasing heights with decreasing observing frequency. The large-scale density distribution, associated with large coronal structures, like streamers and coronal condensations, affects the propagation of radio waves determining global stop and reflecting regions, but the small-scale density distribution of very small structures like the threads can cause reflection and scattering of the radiation, which are in principle detectable through an analysis of directivity and polarization.

Analogously a relation links the local electron gyro-frequency f_{ce} with the local magnetic field intensity B as

$$f_{ce} = f_{ce}(r,\vartheta,\varphi) = 2.80 \cdot B(r,\vartheta,\varphi) \qquad (2)$$

with $f_{ce}(r,\vartheta,\varphi)$ in MHz and $B(r,\vartheta,\varphi)$ in Gauss. Generally in the solar atmosphere cyclotron modes at the fundamental frequency cannot propagate as $f_{ce} < f_{pe}$ (but where the field is quite intense, e.g., above active regions), but if higher order harmonics are generated the propagation condition can be satisfied. If the magnetic field topology is known, e.g., through X-ray observations, and it is possible to ascertain the gyro-radiation mechanism, the magnetic field intensity at the source can be derived [12]. Provided that a coronal density and/or magnetic field model is known, often from optical observations [13, 14], different stages of radio processes can be modelled. Complementarily if the radio emission process (e.g., plasma or cyclotron [12]) is known for a particular radio signature, coronal structures can be mapped in a tomographic way and a coronal density and/or magnetic field model can be derived from radio observations [15]. It appears evident the importance of a reliable electron density distribution and magnetic field topology for a consistent interpretation of solar radio events as well as for an effective 3-D reconstruction of coronal features [11].

The diagnostic capabilities of radio instruments are limited by the available receiving frequencies (size and location of the investigable plasma region) and by the resolution in frequency (spatial resolution in region's radial depth), in space (spatial resolution) and in time (resolution on dynamics of phenomena). Figure 1 reports the wavelength (radial height) coverage of some imaging radio instruments, only a fraction of which is exclusively dedicated to solar research. Figure 2 shows the maximum spatial resolution (at the shortest wavelength) compared to the average size of typical solar features. It is clear that sub-arcsec structures can be hardly resolved using conventional techniques. Moreover it was suggested [19] that synthesis instruments cannot exploit the maximum instrumental resolution at wavelength shorter than a few decimeters due to angular broadening of the brightness distribution on coronal electron density fluctuations, as no fine radio structures were observed in this domain so far. We will concisely comment on this point in Section 4.

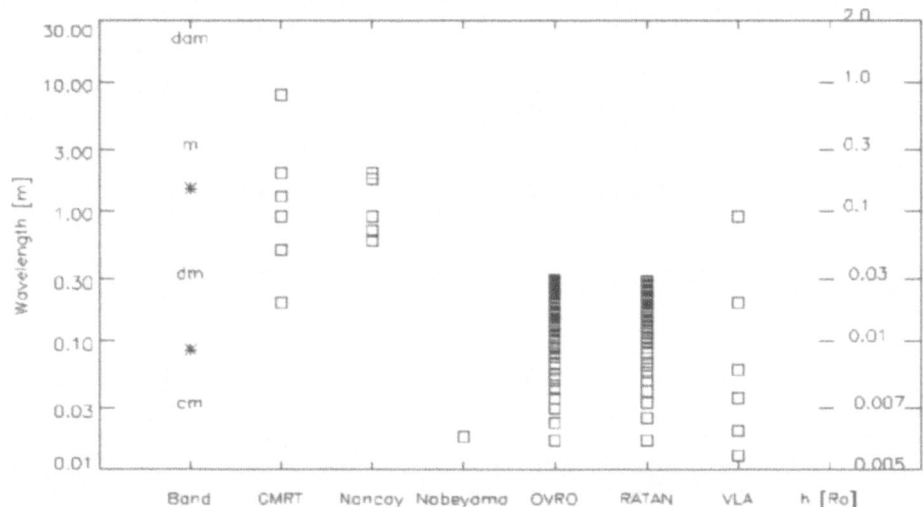

Figure 1. Wavelength coverage of different imaging radio instruments. The approximate corresponding radial distances (solar radii above the photosphere) are indicated in the right ordinate. (GMRT-Giant Meter wavelength Radio telescope, India; Nançay Radioheliograph, France; VLA-Very Large Array, USA; RATAN-600, Russia; OVRO-Owens Valley Radio interferometric system, USA; Nobeyama Radioheliograph, Japan. Data from [18].)

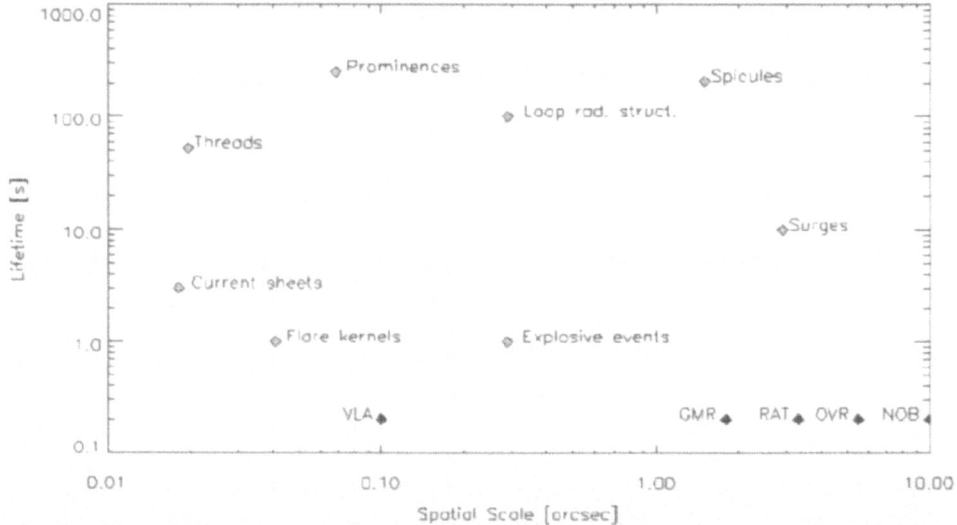

Figure 2. Maximum spatial resolution of imaging radio instruments.

3. Radio observations during solar eclipses

The maximum spatial resolution of a radio telescope is realized at its shortest operating wavelength and depends on the reciprocal of the equivalent aperture of the antenna system. During solar eclipses such instrumental limit can be overcome through the occulting effect of the moon's limb which scans the field of view with the net result of increasing the spatial resolution by an order of magnitude or more with respect to the instrumental one with no sidelobes effect and high positional accuracy. In general the analysis of the slope of the eclipse curve allows to derive the radial brightness distribution of an eclipsed object at any wavelength and for any kind of eclipse[20], whereas the average flux density of thin concentric bands can be determined during total eclipses by measuring the brightness distribution of solar limb [21] (see Table 1).

TABLE 1.   Main methods used in eclipse observations at radio wavelengths

Authors	Measure	λ	Eclipse	Method	Results
Hagen and Swanson [20]	Radial brightness distribution of an eclipsed object	any	any kind	Slope of the eclipse curve	Radial brightness distribution
Smith [21]	Brightness distribution of solar limb	3.3 mm	Total eclipse	Thin concentric bands	Average flux density of bands

The above techniques were used in past eclipses in radio observations mostly at very short wavelengths which map the chromosphere (λ~mm) and the low corona (λ~cm). In this brief review we schematically present a significant subset of the relevant literature in tabular form for reference purposes, organized in four categories according to main scientific targets: 1. Determination of brightness distribution and disk radius (Table 2); 2. Mapping of the quiet Sun (Table 3); 3. High resolution scan of active regions (Table 4); 4. High resolution scan of quiescent prominences (Table 5).

3.1 RADIO BRIGHTNESS DISTRIBUTION AND DISK RADIUS

The brightness distribution over the solar radio disk at different wavelengths is an important diagnostic of temperature and density distribution with height. The size of the radio disk is approximately that of the optical disk in the millimetric band and increases with wavelength. Eclipse observations allow a precise derivation of the brightness distribution and the radius of the disk (Table 2). Limb brightening effects were detected at sub-mm [30] and mm [25, 26, 28] wavelengths, ascribed to the presence of spiculae by some authors [28], but recent millimetric observations [29] did not found any evidences of limb effects as well as a non-eclipse drift scan of the Sun [22]. Similarly no singularities at limb were detected in a centimetric observation [27]. Limb darkening at the North pole was instead present at centimetric wavelengths and interpreted as due to the presence of a coronal hole [23] and to a lower density in the region [24]. Such results, seemingly contradictory, can be explained in terms of appropriate temperature and density models and define important constraints on their derivation. The accurate measurements of radii (1.0048 R_o, 1.0079 R_o, 1.014 R_o at 850 μm, 3 mm, and 2 cm respectively) are consistent with the theoretical estimations.

TABLE 2. Eclipse radio observations of the solar disk at sub-mm, mm and cm wavelengths

Authors	Measure	λ	Eclipse	Method	Relevant results
Joensen et al. [22]	Radial brightness distribution	3.3 mm	Non-eclipse	Derivatives of drift scan of Sun and Moon	Limb neutrality
Drago and Patriarchi [23]	Brightness distribution at N Pole	3 cm 21 cm	Partial eclipse 1971	Eclipse curve	Darkening @ 21 cm suggesting a coronal hole
Kaufmann and Marches [24]	Brightness distribution at N Pole	4.3 cm	Total eclipse 1973	Eclipse curve	Limb darkening suggesting $N_{EQ} \cong 1.4\ N_P$
Swanson and Hagen [25]	Radial brightness distribution	3.2 mm	Total eclipse 1970&73	Eclipse curve	Local limb brightening (15%)
Shimabukuro et al. [26]	Radial brightness distribution	3.3 mm	Total eclipse 1973	Eclipse curve	Local limb brightening (20%) Total of order 2%
Abramov et al. [27]	Radial brightness distribution	2 cm	Partial eclipse 1975	Eclipse curve	No singularities at limb R = 1.014 ± 0.004
Labrum et al. [28]	Radial brightness distribution	3 mm	Total eclipse 1976	Eclipse curve	Slight limb brightening (spicules?) R = 1.007 ± 0.001
Belkora et al. [29]	Solar limb at point of 1st contact	3 mm (1.6")	Total eclipse 1991	Eclipse curve	No bright/dark trend R = 1.0079 ± 0.00083 (5500 km; top of spiculae)
Ewell et at. [30]	Radial brightness distribution	850 μm	Total eclipse 1991	Eclipse curve	Local limb brightening (35%) Total of order 10% R = 1.0048 ± 0.00020

3.2 RADIO MAPPING OF THE QUIET SUN

By combining the eclipse curve with aperture synthesis, Marsh et al. [31] (Table 3) used the VLA to map the quiet Sun at 6 cm with a resolution of 2.5 x 12.7 arcsecs in order to analyze the small scale structure. They showed that the scanned region of the disk is dominated by a small number of radio sources with spatial dimensions 9 to 25 arcsecs and brightness temperatures in the range $(6-8)\cdot10^4$ K, an order of magnitude higher than previously derived. The authors stress the association with bipolar magnetic regions and suggest an interpretation in terms of X-ray bright points.

TABLE 3. Eclipse radio observations of the quiet Sun at cm wavelengths

Authors	Target	λ	Eclipse	Method	Relevant results
Marsh et al. [31]	Quiet Sun	6 cm (VLA 2.5"x12.7" + H_α)	Partial eclipse 1977	Eclipse curve combined with aperture synthesis	Small number of sources ($T_b \sim 8 \times 10^4$ K; $\theta \sim 15$"); possible association with X-ray Bright Points

3.3 RADIO SCAN OF ACTIVE REGIONS

Multi-wavelength radio observations of eclipsed active regions in the cm band map the associated microwave Slowly-Varying (SV) component. The tomography allows the identification and location of radiation mechanisms (e.g., gyroradiation in the loop's legs and at the top and a thermal bremsstrahlung component [32]), the modelling of the 3-D structure of the electron temperature and density, the derivation of the flux density and brightness temperature spectra [33] and information on the magnetic field topology [32] and strength (e.g., an upper limit to the gradient with height [34]). Very hot, sub-arcsec compact radio sources were observed in quiescent active regions [35] suggestive of efficient plasma confinement in a neutral sheet topology.

TABLE 4.   Eclipse radio observations of Active Regions at cm wavelengths

Authors	Target	λ	Eclipse	Method	Relevant results
Luo et al. [32]	AR with bipolar sunspot group	21 cm 8.2 cm 3.2 cm 2 cm	Total eclipse 1980	Eclipse curve	Magnetic loop structure of μ SV source [gyroradiation in legs and top (dominant) + thermal bremsstrahlung]
Ji et al. [33]	"	"	"	"	3-D T_e, N_e, B S and T_b spectra
Gary et al. [34]	AR	30-1.7 cm (OVRO 1.8" + NIXT)	Total eclipse 1991	Eclipse curve	Magnetic field gradient above sunspots ($|\nabla B| \leq 1.5 \times 10^{-6}$ G/cm). Spectra of T_b.
Correia et al. [35]	Quiescent AR	1.35 cm	Total eclipse 1980&90	Eclipse curve	Very hot ($T_b = 10^7$-10^8 K), small ($\theta \leq 1$"; l~700 km) compact sources

3.4 RADIO SCAN OF QUIESCENT PROMINENCES

High resolution scans of the radio counterpart of prominences and filaments (Table 5) provide physical parameters, such as plasma volume emissivity, temperature, density and height, that constrain also the modelling of the optical features.

For instance, the same height was derived for a radio limb prominence, observed in the mm band, and the optical one [36], whereas the radio width was 10 times larger on the average. Furthermore, for the first time a high resolution transverse

TABLE 5.   Eclipse radio observations of prominences/filaments at mm wavelengths

Authors	Target	λ	Eclipse	Method	Relevant results
Apushkinskii et al. [36]	Quiescent limb prominence	8.0 mm	-	Eclipse curve	h = 25000 km T_b = 7000 K Width = 1.5' T_e = 6500-8300 K N_e = 5x10^9 cm^{-3}
Bracewell and Graf [37]	Quiescent filament	9.0 , 3.5, 1.2 mm	1977	Eclipse curve	Emission feature in the cavity

scan of a "radio filament" (usually observed as a brightness depression larger than the optical dark filament) showed a narrow emission feature in the core of the depression and correspondent to the position of the H_α filament [37], favouring the model of a dense filament in a surrounding low-density volume (cavity).

4. Discussion and conclusions

We considered different kind of radio observations carried out during solar eclipses which provided detailed, and often new, information on the spatial structure of the atmospheric plasma with improved resolution, but we also stressed the spatial resolution limits in radio domain imaging. With regard to that we briefly mention some alternative radio diagnostics which provide indirect evidence of small coronal inhomogeneities. In fact, radio waves can be reflected on ordered, denser coronal structures (fibers), which maintains the directivity but increases the apparent source size [38]. Similarly refraction in large-scale overdense structures, e.g. in streamers, affects position and source size, as was proposed for type III radiation from electron streams and the propagation in the fibrous structure of the streamer determines the fine frequency structure of the events [39]. Density inhomogeneities are invoked also to explain the observed bandwidth and duration of type III bursts in the metric-decametric bands [40]. Hence we expect electron stream radiation to be an effective diagnostic [41] of density and magnetic field of fine structures, even with no direct dependence on the maximum instrumental resolution available, assuming a suitable radiation model.

Acknowledgements

The S.O.C. is gratefully acknowledged for the support and the invaluable collaboration during the preparation of this paper as well as the L.O.C. for the excellent hospitality.

References

1. Koutchmy, S. (1994) Coronal physics from eclipse observations, in A.H. Gabriel (ed.), *The Solar Corona, Adv. Sp. Res.* **14**, 29-39.
2. Koutchmy, S. (*this issue*) Fine-scale coronal structures deduced from eclipse observations.
3. Livingston, W.C. (*this issue*) High resolution images of coronal features.
4. Vial, J.C. (1994) The Solar Corona from SOHO, in A.H. Gabriel (ed.), *The Solar Corona, Adv. Sp. Res.* **14**, 181-194.
5. Wenzel, K.-P. and Marsden, R.G. (1994) Observations of the solar corona with Ulysses: expectations above the polar regions, *ibid.*
6. Dame', L., Rutten, R.J., Thorne, A.P., and Vial, J.C. (1994) The SIMURIS interferometric mission: solar physics objectives and model payload, *ibid.*
7. Bely-Dubau, F. (1994) Spectroscopic methods for coronal diagnostics, *ibid.*
8. Clette, F. (*this issue*) Reconstruction of the large-scale distribution of coronal electrons from eclipse data.
9. Hassler, D.M., Gardner, L.D., and Kohl, J.L. (1992) Measuring electron temperature in the extended corona, in V. Domingo, A. Poland and J. Mariska (eds.), *Proc. 1st SOHO Workshop, Annapolis (USA)*, ESA SP-348, 375-377.
10. Dulk, G.A. and McLean, D.J. (1978) Coronal magnetic fields, *Solar Phys.* **57**, 279-295.
11. Aschwanden, M.J. (1995) Imaging, stereoscopy, and tomography of the solar corona in soft X-rays and radio, in A.O. Benz and A. Krueger (eds.), *Coronal Magnetic Energy Releases*, Proc. CESRA Workshop, Caputh (Germany), *Lecture Notes in Phys.* **444**, 13-34.

66

12. Dulk, G.A. (1985) Radio emissions from the Sun and stars, *Ann. Rev. Astron. Astrophys.* **23**, 169-224.
13. Newkirk, G., Jr. (1961) The solar corona in active regions and the thermal origin of the slowly varying component of solar radio radiation, *Ap. J.* **133**, 983-1013.
14. Saito, K., Poland, A.I., and Munro, R.H. (1977) , A study of the background corona near solar minimum, *Solar Phys.* **55**, 121-134.
15. Lang, K.R. (1992) Very Large Array (VLA) observations of coronal loops, in V. Domingo, A. Poland and J. Mariska (eds.), *Proc. 1st SOHO Workshop, Annapolis (USA)*, ESA SP-348, 361-365.
16. Lantos, P. and Alissandrakis, C.E. (1992) Observation of a coronal streamer at meter and decameter wavelengths, *ibid.*, 141-143.
17. Schmal, E.J., Gopalswamy, N., and Kundu, M.R. (1992) Models of coronal streamers at meter-decameter wavelengths, *ibid.*, 145-148.
18. Price, R.M. (1989) Radio astronomy observatories, Committee on Radio Frequencies of National Research Council (CORF), National Academy Press, Washington, D.C.
19. Bastian, T.S. (1994) Angular scattering of solar radio emission by coronal turbulence, *Ap. J.* **426**, 774-781.
20. Hagen, J.P. and Swanson, P.N. (1975) A method for deriving radial brightness distributions from eclipse observations, *Ap. J.* **198**, 219-222.
21. Smith, P.L. (1975) Estimation of solar limb brightening at radio wavelengths from solar eclipse measurements, *IEEE Trans. on Antennas and Propagation* **AP-23**, 237-241.
22. Joensen, P., McCutcheon, W.H., and Shuter, W.L.H. (1974) A method for investigating the brightness distribution near the solar limb at millimeter wavelengths, *Solar Phys.* **39**, 309-314.
23. Chiuderi Drago, F. and Patriarchi, P. (1974) Brightness distribution at $\lambda = 3$ and 21 cm near the solar north pole, *Solar Phys.* **37**, 403-408.
24. Kaufmann, P. and Marques dos Santos, P. (1974) Characteristics of solar north polar cap radio emission at 4.3 cm based on 24 December 1973 eclipse observations, *Astron. Astrophys.* **35**, 459-461.
25. Swanson, P.N. and Hagen, J.P. (1975) The radial brightness distribution of the Sun at 3.2 mm as determined from the June 30, 1973 total solar eclipse and a reanalysis of the March 7, 1970 total solar eclipse, *Solar Phys.* **43**, 57-62.
26. Shimabukuro, F.I., Wilson, W.J., Mori, T.T., and Smith, P.L. (1975) 3.3 millimeter limb brightening measurements during the 30 June 1973 total solar eclipse, *Solar Phys.* **40**, 359-370.
27. Abramov, Iu.P., Avdiushin, S.I., Andrianov, S.A., Barabanshchikov, Iu.F., Drozdovskii, A.A., and Iasnov, L.V. (1978) Preliminary results of observations of the solar eclipse of May 11, 1975, *Radioizluchenie Solntsa* **4**, 158-162.
28. Labrum, N.R., Archer, J.W., and Smith, C.J. (1978) Solar brightness distribution at 3 mm wavelength from observations of the eclipse of 1976 October 23, *Solar Phys.* **59**, 331-343.
29. Belkora, L., Hurford, G.J., Gary, D.E., and Woody, D.P. (1992) Measurement of the solar limb brightness profile at 3 millimeters during the total eclipse of 1991 July 11, *Ap. J.* **400**, 692-698.
30. Ewell, M.W., Jr., Zirin, H., Jensen, J.B., and Bastian, T.S. (1993) Submillimeter observations of the 1991 July 11 total solar eclipse, *Ap. J.* **403**, 426-433.
31. Marsh, K.A., Hurford, G.J., and Zirin, H. (1980) VLA observations of spatial structure in the quiet Sun at 6 centimeters, during the 1977 October eclipse, *Ap. J.* **236**, 1017-1025.
32. Lou, X.H., Yao, D.Y., and Yan, Y.Y. (1982) Solar eclipse observations of a magnetic loop structure at 2, 3.2, 8.2 and 21 cm, *Chinese Astron. Astrophys.* **6**, 261-266.
33. Ji, S., Zhao, R., Zhou, L., and Luo, X. (1993) Combined radiation mechanism in the Sun's Active Region no. 75 during the eclipse of 16 February, 1980, *Astrophys. Sp. Sci.* **200**, 239-250.
34. Correia, E., Kaufmann, P., and Strauss, F.M. (1992) Compact sources of suprathermal microwave emission detected in quiescent active regions during lunar occultations, *Solar Phys.* **138**, 223-231.
35. Gary, D.E., Leblanc, Y., Dulk, G.A., and Golub, L. (1993) The magnetic field in the corona above sunspots at the eclipse of 1991 July 11, *Ap. J.* **412**, 421-430.
36. Apushkinskii, G.P., Berulis, I.I., Losovskii, B.Ia., Sorochenko, R.L., Tsygarov, A.N., and Iasnov, L.V. (1976) Radio-astronomical observations of a prominence during the solar eclipse of May 11, 1975, *Astronomicheskii Zhurnal* **53**, 1249-1253.
37. Bracewell, R.N. and Graf, W. (1981) Solar quiescent prominences at 10.7 GHz, *Nature* **290**, 758-759.
38. Bougeret, J.L. and Steinberg, J.L. (1977) A new scattering process above solar active regions: propagation in a fibrous medium, *Astron. Astrophys.* **61**, 777-783.
39. Itkina, M.A. and Levin, B.N. (1992) A computer simulation study of type III radio burst propagation through the solar corona, *Astron. Astrophys.* **253**, 521-524.
40. Roelof, E.C. and Pick, M. (1989) Type III radio bursts in a fibrous corona, *Astron. Astrophys.* **210**, 417-424.

41. Messerotti, M (1996) Solar beam-plasma radiation: observational diagnostics, in M. Tessarotto (ed.), *Proc. IV Symposium on Plasma Dynamics: Theory and Applications (Trieste, Italy, 1995)*, MHD Consortium of the Trieste University (in the press). Osservatorio Astronomico di Trieste, Preprint n. 217 / 1996.

COMPUTATION OF NONLINEAR FORCE-FREE CORONAL MAGNETIC FIELDS: THEORY FACE TO OBSERVATIONS

C.M. ORZARU

*Astronomical Institute of the Romanian Academy of Sciences,
Str. Cutitul de Argint 5, 75212 Bucharest, Romania*

Abstract. In this paper, various methods used for the extrapolation of the nonlinear force-free coronal magnetic fields from photospheric data are analyzed, and their results (both theoretical and numerical) compared with observational data.

1. Introduction

The striking, very complex forms of evolution of coronal magnetic fields (rooted probably in the dynamics of the sub-photospheric processes), have as the main contributor the magnetic field. Although the soft X-ray images of the corona regularly provided by the Yohkoh satellite, the UV images taken by the Solar Maximum Mission (SMM) instruments improved our view of the Sun's large-scale field, revealing the extremely inhomogeneous structure of the corona, for a deeper insight in its three dimensional structure, we have to rely on the extrapolations of the magnetic field measured in the cooler layers (photosphere, chromosphere and in the prominences). This necessity is motivated by the difficulty of measuring the magnetic field in the corona, because its lines are optically thin (and the line profiles give only an average value along the line-of-sight); another reason is that in this layer the thermal broadening is more important than the Zeeman effect [4]. The results of magnetic field's extrapolation are then compared with observational data.

2. Force-Free Fields: Definition, Derivation, Validity

In the low corona, the governing equation for plasma (in static equilibrium) is:

$$- \nabla p + 1/4\pi \, \mathrm{curl}\, \boldsymbol{B} \times \boldsymbol{B} + \rho \boldsymbol{g} = \boldsymbol{0} \tag{1}$$

Here ∇p and $\boldsymbol{g}$ are negligible in comparison to the other terms, so eq. (1) is equivalent to:

$$\mathrm{curl}\, \boldsymbol{B} = \alpha(\boldsymbol{r})\boldsymbol{B} , \tag{2}$$

Z. Mouradian and M. Stavinschi (eds.),
Theoretical and Observational Problems Related to Solar Eclipses. 69-72.
©1997 *Kluwer Academic Publishers.*

Applying the Stokes theorem to the integral form of eq. (2), the physical significance of α is more evident, as the degree of twist of the field. For $\alpha = 0$, $J = 0$, hence the twist is nul and the field is potential. The eq. (2), together with div $(B) = 0$ are describing the so-called force-free field. These two equations imply:

$$B \cdot \nabla \alpha(r) = 0 \qquad (3)$$

in the domain $D = \{z>0\}$(i.e. above the photospheric level). Eq. (3) states that α is constant along a field line; if α has the same value for all the field lines, the field is *linear*. But the reasons pleading for α=ct. are few: corona's satellite observations in EUV and soft X-ray [11] show that for coronal structures delimitating magnetic fields at different heights, different sets of force-free fields (different values of α) are found.

The force-free field is a state which makes the energy stationary, assuming a given flux distribution and connectivity of force lines [12]. This force-free assumption for the field is motivated by the argument that. in the low corona, the plasma β is much lower than unity (e.g. in prominences $\beta = 8\pi P/B^2 \approx 10^{-2}$) due to the decline of hydrodynamic pressure with height, which is much faster than the one of the magnetic pressure (because of flux divergence); it should be nevertheless mentioned that this approximation is not uniformly valid in the corona (e.g. for fields of about 30 mT in the active regions, the magnetic pressure can exceed 400 Pa, while empirical estimates of the gas pressure in the coronal active regions, range between 0.1-1 Pa.)

But how valid is this statement at the photospheric level and in the chromosphere ?

In the photosphere, B is clearly not force-free, due to the high plasma β (but one should check that the three components of the force and momentum integrated over $\{z=0\}$, vanish; this test is not easy to do in the present, since the measurements are not accurate enough); higher in the atmosphere, this condition becomes more likely to hold.

2.1 NONLINEAR FORCE-FREE FIELDS (NLFF)

Though more realistic (even observed - see .[6],[7]), NLFF (for which α depends of the configuration of B) are more difficult to construct. A particular solution doesn't help to derive a general one (it is also difficult to find for which boundary conditions the problem makes sense).

The structure of eq. (2) and (3) is mixed elliptic-hyperbolic, i.e. both are elliptic for B and (3) is hyperbolic for α. For the elliptic component, one needs $B_{n/\{z=0\}}$, and for the eq. (3) one should provide the value of α for every surface Σ (e.g. on ∂D) which intersects B. The value of α on $\partial D^+ = \{z \in \partial D \mid B_n > 0\}$can be prescribed; problems arise for a field configuration containing not connected subdomains of D, for the values of α not transmitted from ∂D to D.

Historically, the nonlinear problem approach has been done: 1) through analytical 2-D models [11],[9]; 2) using global theorems from the theory of nonlinear PDEs [1]; 3) using numerical techniques.

3. Methods of Computation of NLFF

The main approaches to the nonlinear problem are:

a) _The Cauchy problem_ [17],[3] - which is a mathematically ill-posed problem -: the magnetic field measured in photospheric vector magnetograms is extrapolated upwards, by means of numerical difference schemes, the validity of the method being checked in comparison with analytical solutions. The disadvantages are the problems due to the noise influence, stability of the numerical scheme with height, neutral points.

b) _the current loop minimization method_ [13], uses the possibility to compute the electric current density J_z from vector magnetograms; it is based on the minimization of the Lorentz force on the currents 'looped' initially through the potential field lines, using a discretization method. The scheme is iterated until the value of $J \times B$ approaches 0. The scheme is efficient in the case of a large potential field, with a small number of individual currents; for a more complex configuration, the scheme is very expensive as time and computer resources.

c) _Spectral methods: the method of weighted residuals_ (MWR): the approximate solution of eq. (2)-(3) is represented as:

$$B^N(r) = \sum_{l,m,n} \beta^{l,m,n} \psi_{l,m,n}(r)$$

$$\alpha^N(r) = \sum_{l,m,n} \alpha^{l,m,n} \psi_{l,m,n}(r) \quad ,$$

where $\psi_{l,m,n}(r)$ is a set of trial functions ($\psi_{l,m,n} \in \mathbf{C}^\infty$), β and α are the expansion coefficients and N is the truncation order in the expansion; in general B^N and α^N do not satisfy eq. (2)-(3), so one minimizes the residuals $\mathfrak{R}^1(r, \beta^{l,m,n}, \alpha^{l,m,n}) = curl\, B^N -$ $\alpha^N(r)B$ and also $\mathfrak{R}^2(r, \beta^{l,m,n}) = div(B^N)$, relative to a chosen norm.

d) the _Euler potential relaxation method_ [14], uses the Euler potential form of B and a relaxation method for a set of force-free equations, linear in the second order derivatives of ∇u and ∇v; the field lines are determined, using a mesh discretization, with a 2D representation for u and v.

e) the _variational energy method_ [15],[16], uses also the Euler potential form of B but in 3D, with the purpose of energy minimization, by means of variational techniques; one defines the Euler potentials on a 3D grid and searches iteratively a minimum for the energy, using the conjugate gradient method. Requiring a Lagrangian description for the field lines, this scheme is only partially dynamic.

f) the _integral approach method_ [8] is based on the analytic continuation, using the Cauchy integral formula in order to solve a von Neumann b.v.p. for the force-free equations. It isn't easy to solve it numerically, due to a combination of volume and integral terms, but the author proves the existence and unicity of the solution.

g) the _inverse coordinate method_ [2] uses a more 'natural' system of coordinates (the inverse flux coordinates) to express the set of nonlinear equations. The resulting quasi-linear elliptic system is solved using a full multigrid method, with a smoothing procedure (line relaxation). The good convergence rate and the simplified free-boundary are the method's main advantages.

h) extrapolation based on *prominences magnetic field:* the longitudinal component of the field is measured via the Zeeman effect, while the three components of the field are obtained through the Hanlé effect, using polarization measurements; as the results seem to differ from the photospheric data, some models treat a mixed-boundary problem, considering both measurements .

4. Comparison with Observational Data

Wu *et al.* [18] studied the AR 6659, using photospheric observations (vector magnetograms) from Marshall Space Flight Center (1991) as boundary conditions ; using the method described in [17], they compute the potential field. Then, following the model, they compute the 3D magnetic field configuration (and other parameters). The model proved an accurate enough representation of the field's structures (mostly an observed two-ribbon flare).

Acknowledgments

The author gratefully acknowledges the informations received from Dr. P. Fox about more recent works in the domain and the valuable discussions had with him, Prof. Dr. P. Ulmschneider, Dr. T. Forbes, Dr. Z. Mouradian and Dr. S. Koutchmy.

References :

1. Aly, J.J. (1989), On the Reconstruction of the Nonlinear Force-Free Coronal Magnetic Field from Boundary Data, *Solar Phys.* **120**, 19-48.
2. Cally, P.S., An Inverse Coordinate Multigrid Method for Free Boundary Magnetohydrostatics, *J. Comput. Phys.* **93**, 411-425.
3. Cuperman, S., Ofman, L., and Semel, M. (1990), *Astron. Astrophys.* **230**, 193.
4. Demoulin, P., Raadu, M.A. (1992), Helical Structures Around Quiescent Solar Prominences Computed From Observable Magnetic Fields, *Solar Phys.* **142**, 291-311.
5. Demoulin, P., Henoux, J.C., and Mandrini, C.H. (1992), Development of a topological model for solar flares, *Solar Phys.* **139**, 105-123.
6. Hagyard , M.J. (1988), *Solar Phys.* **115**, 107.
7. Hofmann, A., and Kalman, B. (1991), *Astron. Astrophys.* **241**, 203.
8. Kress, R. (1978), *Proc. Royal Soc. Edinburgh*, **82A**, 71.
9. Low, B.C. (1982), Magnetic Field Configurations Associated with Polarity Intrusion in a Solar Active Region, *Solar Phys.* **77**, 43-61.
10. Low, B.C., and Lou, Y.Q. (1990), Ap.J. **352**, 343.
11 Neupert, W.M., Nakagawa, Y., and Rust, D.M. (1975), *Solar Phys.* **43**, 359.
12. Sakurai, T. (1979), *Publ. Astron. Soc. of Japan* **31**, 209.
13. Sakurai, T. (1981), *Solar Phys.* **69**, 343.
14. Sturrock, P.A., Woodbury, E.T. (1967), in P.A. Sturrock (ed.), *Plasma Astrophysics*, Academic Press, 155.
15. van Ballegooijen, A.A. (1988a), in Proc. Ninth Sacramento Peak Summer Workshop on Solar and Stellar Coronal Structure and Dynamics, ed. R.C. Altrock, 115.
16. van Ballegooijen, A.A. (1988b), *Geophys. Ap. Fluid Dyn.* **41**, 181.
17. Wu, S.T., Sun, M.T., Chang, H.M., Hagyard, M.J., and Gary, G.A. (1990), *Ap.J.* **362**, 698.
18. Wu, S.T., Weng, F.S., Hagyard, M.J., Machado, M., Schmieder, B. (1994), IAU Colloq. 144 *"Solar Coronal Structures"*, V. Rusin, P. Heinzel and J.C. Vial (ed.), 201-205.

COMPARING THE LARGE-SCALE CORONAL ELECTRON DENSITY DISTRIBUTION OF THE 1991 AND 1994 SOLAR ECLIPSES

J.-R. GABRYL, P. CUGNON and F. CLETTE
Observatoire Royal de Belgique
Avenue Circulaire 3, B-1180 Bruxelles, Belgium

Abstract. The observational determination of the electron density and velocity distributions in the solar corona is a basis to any physical modelling. In this context, our program, started in 1973, consists in deriving the large-scale electron density distribution from photographic observations of the polarized white-light corona during total solar eclipses. Since 1991, this program makes use of CCD technologies. We present here results of our 2D descriptive model of the electron density distribution.

1. The 1991 and 1994 Eclipse Observations

Both eclipses were observed with a CCD camera controlled by a portable PC and located at the focus of a 200 mm telelens (11" per pixel). A linear polarizer at the top of the telelens could be rotated over 12 indexed positions 15° apart. The linear polarization information is then four times oversampled. Due to the 8 bit digitization, 3 series of different exposures were necessary to cover the whole range of coronal intensity from the limb to a distance of at least 3 $R_\odot$. The 36 raw images were first realigned using a intercorrelation algorithm, then corrected for the sensor black level and flat-fielded.

The 1991 data reduction leading to the determination of the K-corona total and polarized brightness (K_t+K_r, K_t-K_r) is detailed in [3,4]. The data processing for the 1994 eclipse observed at Putre (Chile) revealed itself more difficult: each image was altered by a varying sky transparency due to moving high-altitude clouds during the totality. Thanks to the oversampling, we were able to apply, at least locally, a linearly varying correction to each image of the 3 series, by reflecting the periodic modulation of polarized brightness measurements (continuity between last and first images). The calculated linear Stokes parameters for each series were then combined. In order to correct for the difference in sky transparency, we forced the observed radial profile of the intensities ratio $F/(K+F)$ in the North polar hole to the theoretical profile extracted from Saito's tables [9]. In order to account for the mean attenuation due to clouds, the total intensity was multiplied by a correction factor to match the theoretical and measured K-corona polarization rates in the same region. Figure 1 shows the total intensity, the polarization rate and the polarized brightness of the 1994 corona. The shape is highly flattened. The polarization rates in the holes are low (10 to 20%), and the highest values (45%) are reached in the West streamer (between 1.4 and 1.5 $R_\odot$).

73

Z. Mouradian and M. Stavinschi (eds.),
Theoretical and Observational Problems Related to Solar Eclipses, 73-76.
© 1997 *Kluwer Academic Publishers.*

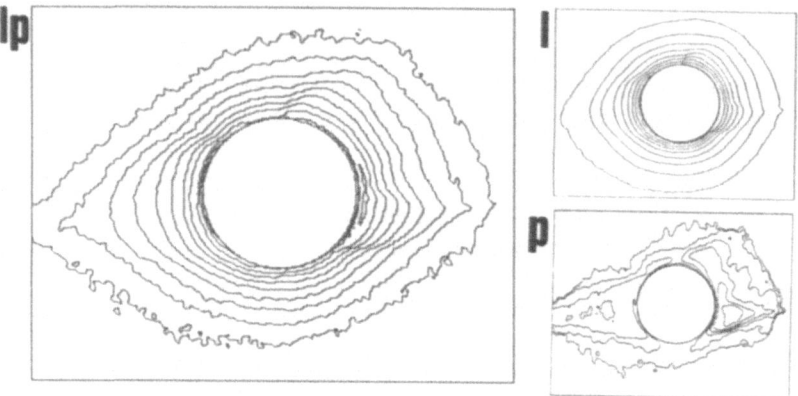

Figure 1. I and Ip: total intensity and polarized brightness in logarithmic scale. The factor between two isophotes is 2. The outer level is 10^{-8} mean $B_\odot$ for I and $2 \cdot 10^{-9}$ mean $B_\odot$ for Ip. The polarization rate p starts at 15 % and the step between two curves is 5%.

2. Modelling the Electron Density Distribution

The 2D information from the eclipse observations of the corona is not sufficient to invert van de Hulst's integral equations. Therefore, an hypothesis on the coronal structure is necessary to rebuild a 3D distribution: we assume an axisymmetric distribution expanded into Legendre polynomials to describe the latitude variations. For each polynomial, a function of 1/r describes the radial decrease in density (Scandariato, Gabryl, Clette, Cugnon model - SGCC [4,5]):

$$N_e(r,\varphi) = \sum_{n=0}^{N_{max}} P_n(\sin\varphi) \sum_{i=1}^{I_{max}(n)} a_{i,n} r^{-b_{l,n}} \tag{1}$$

One can freely choose the position of the symmetry axis and the complexity level of the density description. Thanks to the properties of Legendre polynomials, the density expression plugged into the integral equations gives an analytical solution. The fitting is achieved in two successive steps by a least squares method. The first step is an angular analysis of which the calculated coefficients are then adjusted to radial functions of the type r^{-b} [4,5]. Although the axisymmetric hypothesis seems restrictive, only the electrons close to the plane of the sky contribute significantly to the integration along the line-of-sight. So, we can calculate two distinct models considering that the symmetry axis divides the distribution plane into East and West half-planes.

3. Results

We first calculated models with the Sun rotation axis as symmetry axis ("rotation model"). But, as often suggested [6,7], the coronal material may be essentially distributed into the heliospheric current sheet (HCS) which looks like a thick plane during most of the solar activity cycle. Therefore we also calculated models with the mean bipolar magnetic axis as symmetry axis ("magnetic model"). In both cases, the higher the maximum polynomial order (Nmax) is, the more accurately the angular variations are described. At the order 2 or 3 (figures 2a, 3a), the calculations give the best fit to the smoothest varying structures. At this level, the 2D complexity is the same as in Saito's models [9]. The best fit is reached at the order 6 or 7 (figures 2b, 3b), any further increase of this parameter bringing no significant improvement. This conclusion is the same for both rotation and magnetic models, that's why we only epresented high Nmax magnetic models (figures 2c, 3c).

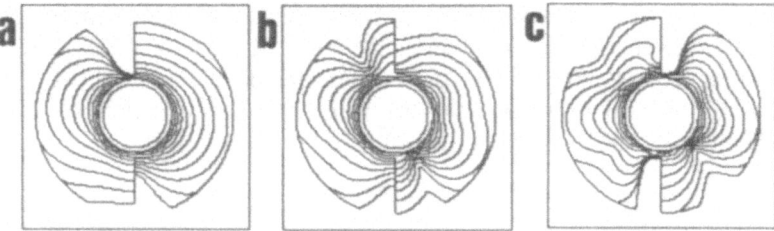

Figure 2. 1994 "rotation models" with Nmax=3 (a) and 7 (b). 1994 "magnetic model" with Nmax=7. The logarithmic curves are spaced by a factor 2, and the external curve corresponds to 1011 e/m3.

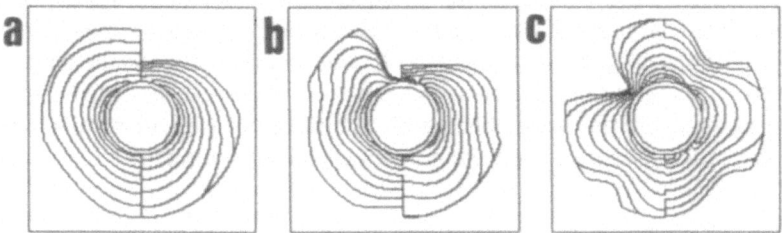

Figure 3. 1991 "rotation models" with Nmax=3 (a) and 6 (b). 1991 "magnetic model" with Nmax=6. The logarithmic curves are spaced by a factor 2, and the external curve corresponds to 4þ1011 e/m3.

High densities are closer to the equatorial plane in 1994 (figure 2) than in 1991 (figure 3). Also, the 1994 densities are 2 to 3 times weaker than in 1991. This may suggest that the density distribution vary by a factor 2 or 3 during the solar cycle. However, we noticed that the 1991 magnetic models are much more East-West symmetric than the similar rotation ones: the electrons seem mainly distributed close to the HCS and along the magnetic axis (figure 3c), in agreement with magnetohydrostatic models [1]. However, the improvement in symmetry is less obvious in 1994 magnetic models: the electrons are present into the HCS, but the density determination in the magnetic poles probably suffers from cloud corrections and calibration errors, so that we miss good values at the poles (figure 2c). The small differences between the 1994 rotation and magnetic models can be explained also by the small angular separation between both axis: 13ø (Gulyaev, 1995) against 64ø in 1991 (Gulyaev, 1992). But in general, the fit in polar regions is poorer due to a intrinsic weaker weightning due to Legendre polynomials.

We also extracted density radial profiles in caracteristic regions (i.e. streamers, holes) from the highest polynomial order models. Our 1991 profiles, even the hole profiles, exhibit values between typical equatorial and streamer values from the compiled tables of Koutchmy [8]. This is probably due to a projection effect from streamers along the line of sight. On the contrary, the 1994 profiles are much closer to the typical mean values. As seen in the 2D maps, the 1991 streamer profiles are more than two times greater than the 1994 ones. Even if our model poorly fits polar densities, our profiles appear consistent with the mean tabulated values.

We finally tried to connect our results with a physical description due to Badalyan [2]. A good fit was generally obtained with her hydrostatic law, but it appears better in 1994 than in 1991 probably due to a superposition of several streamers at different temperatures along the lines of sight in 1991. The deduced temperatures are consistent with the typical plasma temperatures: 1.29 MK (PA=264°), 1.59 MK (PA=99°) and 1.00 MK (PA=201°) in 1994; 1.84 MK (PA=43°), 1.71 MK (PA=219°) and 1.63 MK (PA=150°) in 1991.

4. Conclusions

We have built a powerful tool to describe the large-scale electron density distribution in two dimensions. The best and more detailed representation is obtained with a maximum polynomial order of 6 or 7, but the choice of the symmetry axis is important. The choice of the mean bipolar magnetic field axis as symmetry axis better justify the axisymmetry hypothesis than the choice of the rotation axis. It has also a immediate physical meaning. Despite the mathematical complexity of Ne, our results agree with simple hydrostatic profiles and give consistent plasma temperatures.

References
[1] Bagenal F. and Gibson S. (1991), Modeling the large-scale structure of the solar corona, *J. Geoph. Res.* **96**, NøA10, 17663-17674.
[2] Badalyan O.G. (1986), Polarization of white-light corona under hydrostatic density distribution, *Astron. Astroph.* **169**, 305-312.
[3] Clette F. and Gabryl J.-R. (1992), The July 11, 1991 Solar corona observed in polarized white light. Preliminary results, Proc. 1st SOHO Workshop *ESA-SP* **348**, (ed.) C.Mattok, 351-353.
[4] Clette F., Cugnon P. and Gabryl J.-R. (1994), Modelling the electron density distribution in the July 1991 solar corona, Proc. IAU Coll. 144 *"Solar Coronal Structures"*, eds Rusin, Heinzel and Vial, 535-539.
[5] Clette F., Reconstruction of the large-scale distribution of coronal electrons from eclipse data, this issue.
[6] Gulyaev R.A. (1992), Eclipse observations of the outer solar corona, *Proc. 1st SOHO Workshop* ESA-SP **348**, (ed.) C.Mattok, 133-136.
[7] Gulyaev R.A. and Molodensky M.M. (1995), The shape of the solar corona on November 3, 1994:Prediction and observation, *Rev. Acad. Nacion. Ciencias de Bolivia* **69**, 11-17.
[8] Koutchmy S. (1992), Allen - third edition (1972). Revised by Serge Koutchmy (1992), *Inst. Astroph.Paris, Pr,-publication* Nø404.
[9] Saito K. (1970), A non-spherical axisymmetric model of the solar K corona of the minimum type, *Ann.Tokyo Astron. Obs.*, Second Ser. **12**, 53-120.

2D MHD NUMERICAL MODEL FOR THE STUDY OF DYNAMIC AND MAGNETIC STRUCTURES IN THE SOLAR ATMOSPHERE

M. S. GHIZARU

Astronomical Institute of the Romanian Academy
Cutitul de Argint 5, 75212 Bucharest 28, Romania

Abstract. The need for a numerical processing is first presented as it applies to the study of MHD wave behaviour at the interface between the solar corona and a prominence. The expression of a wave invariant was obtained for a linear model, indicating that a nonlinear approach is necessary to explain the over-reflection mechanisms. A numerical MHD model is then developed, based on the resistive MHD equations written in cylindrical (r,z) coordinates. The numerical integration procedure and the results of a number of tests with various advection schemes are presented in the last section.

1. Introduction

One of the theories for explaining the heating of the solar corona uses MHD waves as main ingredient. Among other physical processes involved in the coronal heating mechanisms, the over-reflection of MHD waves incident on critical layers, might play an important role. These are characterised by strong shear in the magnetic and flow fields. Such a layer might be the interface between the solar corona and a prominence.

MHD waves incident on this layer are normally partially absorbed and partially reflected. In a series of papers, Sakurai, Goossens and Hollweg [1],[2], Goossens, Hollweg and Sakurai [3], Goossens and Hollweg [4], have studied the resonant behaviour of MHD waves on magnetic flux tubes, developing a procedure which yields connection formulae through a resonant surface in 1D equilibrium states. But it might be possible, unter certain special conditions, that the MHD wave, instead of being absorbed inside the flux tube, might extract energy from the mean flow and/or magnetic field in the presence of strong shear, and is over-reflected. To investigate these conditions and the mechanisms by which over-reflection occurs, we have applied the procedure developed by Eltayeb [5] for the linear model from [3], and found an expression for the wave invariant, as presented in the second section below.

As a result, a nonlinear approach became necessary, and this requires a numerical simulation. Towards this end, a 2D model in cylindrical geometry is presented in the last section of this paper. It is based on the MHD resistive equations, and is integrated on an Arakawa b grid, as shown in the paper of Messinger and Arakawa [6]. The operator splitting is used as in Bowers and Wilson [7], and the div**B**=0 constraint is satisfied by deriving the magnetic field from a vector potential. Since the most sensitive point in such a numerical model is the treatment of the

Z. Mouradian and M. Stavinschi (eds.),
Theoretical and Observational Problems Related to Solar Eclipses, 77-80.
© 1997 *Kluwer Academic Publishers.*

advection terms, the result of a few tests using various advection schemes is presented at the end of the paper.

2. Over-Reflection of MHD Waves on Flux Tubes

In their paper, Goossens et al. [3] obtained the equation for the radial component of the Lagrangian displacement in the following form:

$$\frac{d}{dr}\left(f(r,\omega^2)\frac{d}{dr}(r\xi_r)\right) - r(r,\omega^2)r\xi_r = 0$$

I have used the same notation as in [3].

By simple transformations, I wrote the equation in the form used by Eltayeb [5] as a starting point for finding a wave invariant:

$$a(r)\frac{d^2\xi_r}{dr^2} + 2b(r)\frac{d\xi_r}{dr} + c_1(r)\xi_r = 0.$$

It can be shown that the equation can be rewritten as:

$$\frac{d^2\psi}{dr^2} + c\psi = 0,$$

with

$$c = \frac{a(c_1 - db/dr) + b\, da/dr - b^2}{a^2},$$

and following the procedure described in [5], I have found for the wave invariant the expression:

$$Inv = \text{Im}\left(r\frac{D}{C_2}\xi_r^* \frac{d\xi_r}{dr}\right),$$

with D and C_2 having the same meaning as in [3].

Since the invariant is zero on one side of the singularity, the linear case is irrelevant for explaining the detailed mechanism of the over-reflection and a nonlinear approach, treated numerically, is necessary.

3. 2D MHD Numerical Model

The equations used for building the numerical model are the following:

$$\frac{\partial U}{\partial t} = -\frac{1}{r}\frac{\partial (rvU)}{\partial r} - \frac{\partial (uU)}{\partial z} - \frac{\partial p}{\partial z} - J_\theta B_r + J_r B_\theta$$

$$\frac{\partial V}{\partial t} = -\frac{1}{r}\frac{\partial (rvV)}{\partial r} - \frac{\partial (uV)}{\partial z} + \frac{\rho w^2}{r} - \frac{\partial p}{\partial r} + J_\theta B_z - J_z B_\theta$$

$$\frac{\partial W}{\partial t} = -\frac{\partial (rvW)}{\partial r} - \frac{\partial (uW)}{\partial z} - \frac{\rho vw}{r} + J_z B_r - J_r B_z$$

$$\frac{\partial \rho}{\partial t} = -\frac{1}{r}\frac{\partial (rv\rho)}{\partial r} - \frac{\partial (u\rho)}{\partial z}$$

$$\frac{\partial \rho \varepsilon}{\nabla t} = -\frac{1}{r}\frac{\partial (rv\rho\varepsilon)}{\partial r} - \frac{\partial (u\rho\varepsilon)}{\partial z} - p\left(\frac{1}{r}\frac{\partial v}{\partial r} + \frac{\partial u}{\partial z}\right) + \frac{1}{\sigma}\left(J_r^2 + J_z^2 + J_\theta^2\right)$$

$$\frac{\partial B_\theta}{\partial t} = \frac{\partial}{\partial r}\left(\frac{\eta}{r}\frac{\partial (rB_\theta)}{\partial r}\right) + \frac{\partial}{\partial z}\left(\frac{\eta}{r}\frac{\partial (rB_\theta)}{\partial z}\right) - \frac{\partial}{\partial r}(vB_\theta) - \frac{\partial}{\partial z}(uB_\theta) + \frac{\partial}{\partial r}(wB_r) + \frac{\partial}{\partial z}(wB_z)$$

$$\frac{\partial A}{\partial t} = \eta\left(\frac{\partial}{\partial r}\left(\frac{1}{r}\frac{\partial (rA_\theta)}{\partial r}\right) + \frac{\partial}{\partial z}\left(\frac{1}{r}\frac{\partial (rA_\theta)}{\partial z}\right)\right) - \frac{u}{r}\frac{\partial (rA_\theta)}{\partial z} - \frac{v}{r}\frac{\partial (rA_\theta)}{\partial r}$$

$$p = (\gamma - 1)\rho\varepsilon$$

The symbols used in these equations have their usual meaning:
(u,v,w) - the velocity components, (J_z, J_r, J_θ) - the current density components and (B_z, B_r, B_θ) - the magnetic field components in the z, r and θ directions, ρ - the density, ε - the internal energy per unit mass, p - the pressure, γ - the adiabatic exponent, σ - the electrical conductivity, and η - the magnetic diffusivity.

The magnetic vector potential is defined so that the poloidal components of the magnetic field are:

$$B_r = -\frac{\partial A_\theta}{\partial z} = -\frac{1}{r}\frac{\partial (rA_\theta)}{\partial z}$$

$$B_z = \frac{1}{r}\frac{\partial (rA_\theta)}{\partial r}.$$

By making this choice, the $\nabla \cdot \mathbf{B} = 0$ constraint is automatically satisfied in the present model. The current density is derived from the magnetic field as in [7].

I have also used the following notation: $U = \rho u$, $V = \rho v$ and $W = \rho w$.

A staggered grid in both time and space is used for a finite difference representation, as indicated in [7]. The velocity components and the vector potential are placed in the cell corners while the thermodynamic variables and the torroidal component of the magnetic field are defined in the cell centres.

The numerical code is structured in two parts: a hydrodynamic and a magnetic driver. In the hydrodynamic driver, the pressure gradient is first computed, then the mechanical work contribution in the energy equation, and then the mass, energy, and momentum advection. The operator splitting method is used for the contribution of each term in the equations.

The magnetic field driver consists of two parts corresponding to the toroidal and poloidal components, each including the axial and radial diffusion, the Joule heating, and the Lorentz force contribution in the momentum equation. The magnetic field diffusion is treated with an implicit scheme using Gaussian elimination, as in [7].

Special attention was paid to the advection scheme. Various experiments were made using a simple second-order scheme with or without alternating direction. In the rotating cone test, large amplitude oscillations develop after two rotations (1028 iterations). A second order scheme was then applied, to maintain the positive profile, but this flattened it quite severely. Some tests were then made in which the two schemes were combined. The optimum compromise were obtained by applying the simpler second order scheme with alternating directions included, and using the positive monotone-inducing scheme every five time steps, as is graphically presented in the paper by Ghizaru and Goossens [8].

Further numerical experiments will use the model presented here for systematic study of the MHD wave over-reflection problem as well as for investigating polar plume dynamics.

References

1. Sakurai, T., Goossens, M. and Hollweg, J.V. (1991) Resonant behaviour of MHD waves on magnetic flux tubes; I. Connection formulae at the resonant surfaces, *Solar Phys.* **133**, 227-245.
2. Sakurai, T., Goossens, M. and Hollweg, J.V. (1991) Resonant behaviour of MHD waves on magnetic flux tubes; II. Absorbtion of sound waves by sunspots, *Solar Phys.* **133**, 245-262.
3. Goossens, M., Hollweg, J.V. and Sakurai, T. (1992) Resonant behaviour of MHD waves on magnetic flux tubes; III. Effect of equilibrium flow, *Solar Phys.* **138**, 233-255.
4. Goossens, M. and Hollweg, J.V. (1993) Resonant behaviour of MHD waves on magnetic flux tubes; IV. Total resonant absorbtion and MHD radiating eigenmodes, *Solar Phys.* **145**, 19-44.
5. Eltayeb, I.A. (1977) On linear wave motion in magnetic-velocity shears, *Phil. Trans. Roy. Soc. London.* **285** A, 607-636.
6. Messinger, F. and Arakawa, A. (1976) *Numerical Methods used inAatmospheric Mmodels,* GARP Publication series No. 17.
7. Bowers, R.L. and Wilson, J.R. (1991) *Numerical Modeling in Applied Physics and Astrophysics,* Jones and Bartlett Publishers, Boston.
8. Ghizaru, M.S. and Goossens, M.(1996) Numerical model for the study of wave-mean flow interactions in solar magnetohydrodynamics, *Romanian Astr. J.* (in press)

MHD TURBULENCE IN THE SOLAR CORONA

A. MANGENEY
Observatoire de Paris-Meudon,
F-92195 Meudon CEDEX, France

There is now a general agreement about the fact that the energy which is fed into the corona comes, in fine, from the underlying convective zone.

The mainly horizontal motions of the photospheric plasma advect the magnetic lines of force; these motions provide an energy input into the upper layers at a rate sufficient to explain the heating of the corona in these regions [4].

In the regions of open magnetic field lines, photospheric motions are likely to excite essentially wave motions which hopefully are able to dissipate where heat input is necessary.

On the other hand, in regions of the solar corona where the magnetic lines of force are closed, as for example above photospheric active regionss, there are several possible scenarios: either the magnetic configuration evolve slowly through a series of quasistatic equilibria, slowly dissipating the magnetic free energy deposited by the photspheric forcing, or some form of M.H.D. turbulence is maintained with plasma velocity and magnetic field fluctuations covering a wide range of time and spatial scales. For a recent review of these issues, the reader is referred to Narain and Ulmschneider, [3].

I shall consider here some consequences of the "M.H.D. turbulence" scenario. The energy injected by the photospheric motions cascade from the scales characteristic of these photospheric motions, say a few 1000 km, towards dissipative lengths which are very small, say of the order of 1m, due to the very high electrical conductivity of the coronal plasma.

Is such a cascade possible? As emphasized by Van Ballegooijen, [9], the fact that the magnetic field lines are frozen in the photospheric plasma imposes on the evolution of the MHD turbulence a constraint which may prevent the formation of "singularities", i.e. strong currents at scales where dissipative processes may proceed in a reasonnable time scale.

This question is yet not entirely settled, see for example the recent works by Longcope and Strauss, [2], or Einaudi et al.,[1].

We shall assume here, that energy is indeed fed into the dissipative scales, l, the corresponding flux - at the base of the corona - being designed by F. Then, the important point I want to stress is that at those scales, $\sim l$, the coronal plasma is at most weakly collisionnal.

Z. Mouradian and M. Stavinschi (eds.),
Theoretical and Observational Problems Related to Solar Eclipses, 81-84.
© 1997 *Kluwer Academic Publishers.*

The argument goes as follows. Let δB_l be the magnetic fluctuation at scale l, λ the magnetic diffusivity $\lambda = c^2/4\sigma$ in c.g.s. units, σ being the electric conductivity of the coronal plasma assumed here to be constant, and c the velocity of light. Then the dissipation rate (in *ergs cm^{-3} sec^{-1}*) can be estimated by

$$\varepsilon \cong \lambda \frac{\delta B^2}{l^2} f$$

I have introduced a filling factor f for the regions where dissipation occurs to take into account the possible intermittency of the dissipation.

Consider a region of horizontal surface S and height H, the total magnetic energy dissipated is $\varepsilon S H$ and must be equal to the energy input, $F S$. Then

$$\varepsilon = F / B$$

On the other hand, in the vicinity of the dissipative regions, the electric field E_d is essentially determined by the resistive terms in Ohm's law

$$\mathbf{E_d} = \mathbf{J}/\sigma - \mathbf{v} \times \mathbf{B}/c$$

the first term of the right hand side being much larger than the second one which is the electric field due to the advection of the magnetic field by the plasma. Therefore E_d is given, at least in order of magnitude, by

$$E_d \cong \frac{J}{\sigma} \cong \frac{\lambda}{c} \frac{\delta B_l}{l}$$

so that $\varepsilon = f c^2 E^2_d / \lambda$, i.e.,

$$E_d \cong \left(\frac{\lambda \varepsilon}{f c^2} \right)^{1/2}$$

An electron of speed v is submitted through collisions with other charged particles to a collisional drag $F_{drag} = e E_D (v/v_e)^2$, [6] where E_D is the Dreicer field

$$E_D = \frac{e \ln(\Lambda)}{l^2_D}$$

l_D the Debye length, $l_D = v_e/\omega_{pe}$, ω_{pe} being the electron plasma frequecy, v_e the electron thermal velocity and $\ln(\Lambda)$ the Coulomb logarithm which is roughly equal to 20 in the conditions considered here.

We shall now see that in a region of dissipation, the whole electron population is submitted to an electric force greater than the collisional drag, since

$$E_D \cong 8.8 \cdot 10^{-6} \frac{n_{10} \ln(\Lambda)}{T_{ev}} \, cgs$$

where the density n_{10} is measured in units of 10^{10} cm^{-3} and the temperature T_{ev} is measured in electron volts, while

$$E_d \cong 34 \left(\frac{\varepsilon}{f}\right)^{1/2} \left(\frac{n_{10} \ln(\Lambda)}{T_{ev}^{3/2}}\right)^{1/2}$$

Let us apply these relations for a heating flux $F \sim 10^8$ *erg cm^{-2} sec^{-1}* a region of height $H=1000$ *km*, a filling factor for the dissipation regions $f \sim 0.1$, a density of $n \sim 10^9$ and a temperature of 10 *ev*; then $E_D \sim 10^{-6}$ *cgs* while $E_D \sim 10^2$ *cgs*.

These numbers are only indicative; however it is clear that E_d is larger than the Dreicer electric field by several orders of magnitude.

The conclusion then is that dissipative processes are essentially collisionless and imply the acceleration of a large number of charged particles; I have considered here only the case of electrons but for electric fields so much larger than the Dreicer field there will also be a strong acceleration of protons and other charged particles. Note furthermore, that the scalelentgh l where dissipative processes occur does not appear in the estimate of E_d given above.

If the turbulent cascade invoked above really occurs at some height in the solar atmosphere, it is to be expected that in this region the particle distributions will be highly non-thermal with high energy tails which may be compatible with the observations (see for example, Zirin, [11] or Roussell-Dupre,[5]).

We can even speculate that this is what happens in the transition region, where at lot of activity at small scales seems to be present at all times. Part of the suprathermal particles accelerated in the dissipative regions may be allowed to drift accross the magnetic field lines, or the field lines themselves may diffuse into the corona (see for example Zimbardo et al.,[10]). Then they could form the population which by velocity filtration constitute the high temperature corona, as advocated by Scudder,[7], [8].

Many problems remain however to solve before such a scenario could be put on a sound physical basis. Observations which would provide a more detailed knowledge of the non-thermal behaviour of the particle distribution functions are certainly of importance in that respect.

84

References

[1]. Einaudi G., Velli M., Politano H., Pouquet A., (1996), *Astrophys. J.*, **457**, L113
[2]. Longcope D.W., Strauss H.R., (1994), *Astrophys. J.*, **426**, 742
[3]. Narain U., Ulmschneider P., (1996), *Space Sci. Rev.*, **75**, 453
[4]. Parker E. N., (1972), *Astrophys J.*, **174**, 499
[5]. Roussel-Dupre R., (1974), *Solar Phys.*, **68**, 243
[6]. Spitzer L., (1962), *Physics of Fully Ionized Gases,* Interscience, New York
[7]. Scudder J.D., (1992), *Astrophys J.*, **398**, 319
[8]. Scudder J.D., (1994), *Astrophys J.*, **427**, 446
[9]. Van Ballegooijen A.A., (1985), *Astrophys. J* **298**, 421
[10]. Zimbardo G., Veltri L., Basile G., Principato S., (1995), Physics of Plasma, **2**, 2653
[11]. Zirin H., (1968), *Astrophys. J.*, **154**, 799

POST-FLARE LOOPS IN THE CHROMOSPHERE AND CORONA

L. VAN DRIEL-GESZTELYI

(1) Konkoly Observatory, Budapest, Pf. 67, H-1525 Hungary

(2) Observatoire de Paris, Section de Meudon, DASOP, URA 2080 (CNRS) F-92195 Meudon Cedex, France

J.E. WIIK

(1) Institute of Theoretical Astrophysics, University of Oslo, N-0315 Blindern, Oslo 3, Norway

(2) Observatoire de la Cote d'Azur, 06304 Nice Cedex-4, France

AND

B. SCHMIEDER

Observatoire de Paris, Section de Meudon, DASOP, URA 2080 (CNRS) F-92195 Meudon Cedex, France

Abstract. Post-flare loops, which may last for several hours during the gradual phase of powerful flares, can provide spectacular limb events during solar eclipses. According to models, on-going magnetic reconnection keeps creating new loops at an ever-increasing altitude. The newly formed hot loops can cool down to chromospheric temperatures quickly, in a few minutes, or slowly, over a few hours, depending on their density. A good test of the model is to measure the altitude difference between loops observed at different wavelengths which show the loops at different stages of the cooling process. Studying a unique, longest ever, series of X-ray $(6 - 7 \times 10^6$ K) and Hα $(1.5 \times 10^4$ K) data on the X3.9 flare which occurred on June 25, 1992 at 20:11 UT we found similar shape and configuration for hot and cool loops during the entire observing period, and thus confirmed that cool loops indeed evolve from hot loops. On the other hand, we find that the computed cooling times are somewhat shorter than the cooling times we deduct from the relative altitudes of hot and cool loops. This discrepancy may imply that loops do not stay at the same altitude but shrink during the cooling process. Whether this is true or not could conceivably be answered through an analysis of multiwavelength post-flare loop observations, perhaps as a joined space (SOHO, YOHKOH) and ground-based observing effort during the 1999 total eclipse.

Z. Mouradian and M. Stavinschi (eds.),
Theoretical and Observational Problems Related to Solar Eclipses, 85-93.
©1997 *Kluwer Academic Publishers.*

1. Introduction

Post-flare loops have long been observed in Hα after powerful long-duration two-ribbon flares connecting the bright flare ribbons and accompanied by long-lasting coronal soft X-ray brightenings (later resolved as loops), spanning above the magnetic inversion line (see reviews by Švestka [1], Schmieder [2] and references therein). Both hot and cool loops can be observed for several hours and they show considerable expansion during their existence. Since the cooling time of hot X-ray loops is much shorter than their lifetime, it is obvious that some kind of energy release makes their long existence possible. Such loops are now widely believed to be the physical evidence of on-going magnetic field line reconnection during the gradual phase of flares, which keeps forming new loops at an ever-increasing altitude [3], [4], [5], [6]. The hot loops then cool down to appear eventually as Hα loops, but by that time newer hot loops are created at higher altitudes; therefore, at any given time, hot loops should be observed at higher altitudes than the relevant cool loops. Several observations support this scenario e.g. [7], [8], [9], [10], [11], [12], while [13] challenges it.

Using coordinated observations of a large system of post-flare loops on 25-26 June 1992, which produced the longest ever parallel observations taken in X-rays and in Hα of such events, we studied the formation and dynamics of the loops, the relative altitudes of hot and cool loops.

2. Expansion Speed and Relative Altitude of Hot and Cool Loops

A post-flare loop system was observed with the *Yohkoh*/SXT (after the large X-class flare of June 25 at 20:11 UT) for more than 15 hours, and from ground based observatories in Japan (Hida Observatory) and Europe (Pic du Midi, France; Valašské Meziřici, Czech Rep.; Wroclaw, Poland; La Palma, Spain, by the Lockheed group) for more than 19 hours. Unfortunately, Yohkoh missed the first three hours of the flare, but covered the rest of the gradual phase. On the other hand Hα observations started right after the impulsive phase at Hida Observatory (Figure 1), so we could follow the fast early evolution of the loop system. During the entire observing period we see an arcade of loops and are able to follow the evolution of three main loops, which appear to be very similar in Hα and in X-rays (Figures 2, 3).

In order to determine the real expansion speed of the X-ray loops, we measured the positions of the footpoints and the loop tops, and subsequently, after an analysis of the magnetic configuration, reconstructed the true shape of the loops (for the method, see [14]), taking into account effects of the solar rotation. The altitude measurements were made independently for the thin and thick Al filters and gave nearly the same altitudes (Figure 4 a). Since X-ray loops are faint at their footpoints, the precision of these

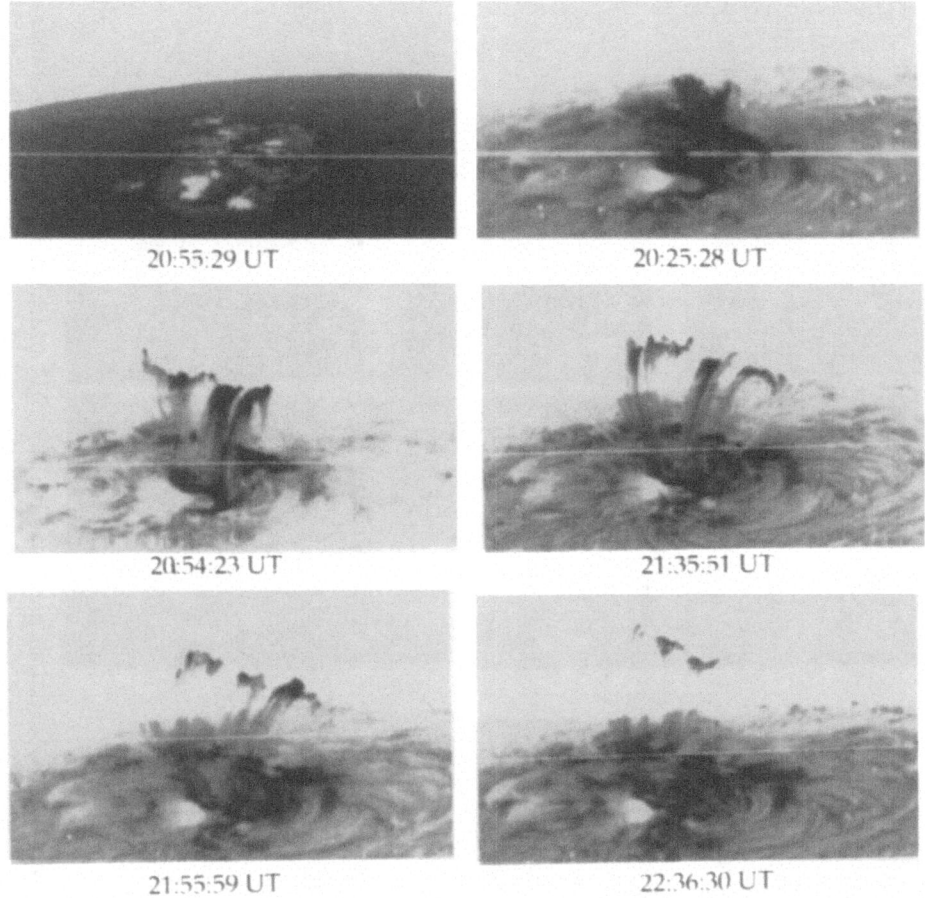

20:55:29 UT   20:25:28 UT

20:54:23 UT   21:35:51 UT

21:55:59 UT   22:36:30 UT

Figure 1. Observations of post-flare loops on June 25, 1992 in Hα at Hida observatory (Japan) during the first three hours of the gradual phase of the flare

measurements decreased when brightenings occurred in the active region. We obtained an almost constant expansion speed of 1.1 km s^{-1} between 02:00 UT and 15:00 UT. Unfortunately, during the first two Yohkoh orbits the footpoints of X-ray loops were outside of the observing field of view, which shortens the dataset for such complete measurements. Since in the Hα images the loop legs are (i) covered by an occulting disc, or (ii) hardly visible on the bright disc, such true height reconstruction was carried out for the X-ray observations only.

To obtain the relative altitudes of the X-ray and Hα loops using the entire data set we had to measure the altitudes of loops from the solar limb, which could be located in all kinds of observations, since the flare in question was a limb event. With error bars due to the determination

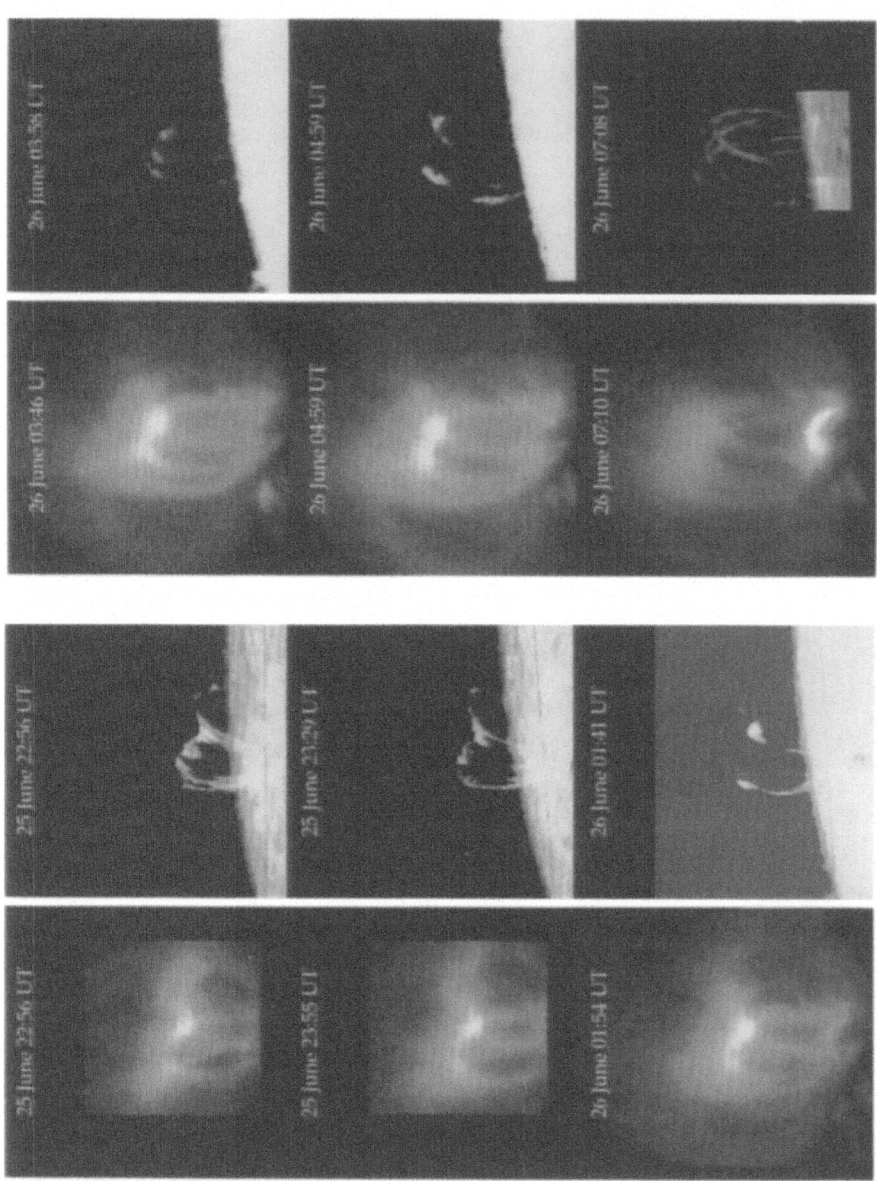

Figure 2. Evolution of the post-flare loops system in soft X-rays (left panels) and in Hα (right panels) during the gradual phase of the X3.9 flare of 25 June 1992 20:11 UT. Note that the Hα and X-ray pictures have the same scale and they are co-aligned.

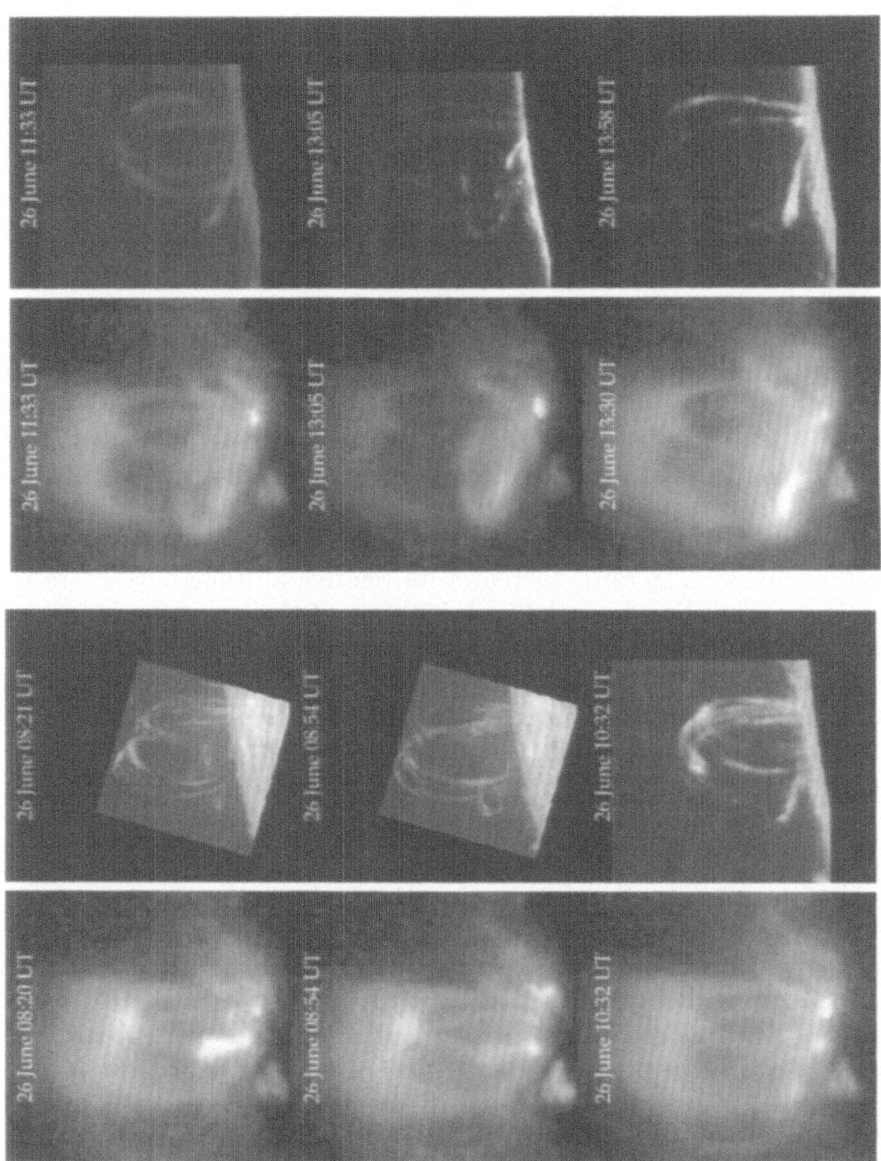

Figure 3. Same as Figure 2., for the late gradual phase of the flare.

of the pixel size of Hα data (around 10%) and to the exposure time of different pictures (in overexposed frames the the solar limb appears to be more extended) we find that hot and cool loops are growing continuously (Figure 4). From the Hα (Hida Observatory) observations, which started 15 minutes after the beginning of the flare (c.f Figure 1), it is clear that the expansion of the loop system was very fast at first ($\simeq$ 8-10 km s^{-1}), it decreased more-or-less exponentially to about 1 km s^{-1} in about four hours, and the loop system was observed to expand with that speed for at least 14 hours.

The three loop systems we can follow seem to show differences in the relative height of the hot and cool loops as they evolve, implying differences in cooling times. This can be due to e.g. (i) differences in their electron density, (ii) differences in the visibility of the loops, due to differences in the intensity of condensation (or blob) formation. In fact the relative visibility of the three loop systems showed changes with time. The loops were not always well defined, especially loop 1 (the smallest loop) was sometimes difficult to follow in X-rays, therefore its relative altitude curves are the least certain. The changing relative brightness of the loops can be seen in Figures 2 and 3), where, e.g., the middle loop (loop 2) is much less bright in X-rays at 08:20 UT than at 04:59 UT. Such changes in X-ray brightness appear in Hα later, clearly showing that cool loops descend from hot loops.

3. Discussion and Conclusions

The expansion rate of the post-flare loop system was found to be of the order of 10 km s^{-1} shortly after the impulsive phase, then to decrease exponentially, and to have stabilized about four hours later at 1.1 km s^{-1}. The shape and structure of the loops are very similar in X-rays and in Hα during the entire gradual phase, considering the fact that the Hα are decendants of X-ray loops observed earlier and not related to the X-ray loops observed at the same time (Figures 2, 3). On the other hand, we have to note that comparing pictures taken at different wavelengths, with different spatial and temporal resolutions is not always an easy task, as loops of the post-flare loop arcade may overlap, or new flares and ejections may occur in the same active region during the long gradual phase. Especially close to the limb, spatially different features may appear projected in superposition, and physically unrelated events can thus be confused.

In [11], using the filter-ratio method for the two Al filters, we determined the temperature and the emission measure EM of the X-ray loops ($n_e^{hot} \simeq$ $7 - 8 \times 10^{10} - 10^{11}$ cm^{-3} around 23:00 UT and $n_e^{hot} \simeq 6 - 7 \times 10^9$ cm^{-3} around 13:00 UT) for the entire gradual phase and computed the electron densities of the hot plasma, taking 4000 km for the line-of-sight thickness

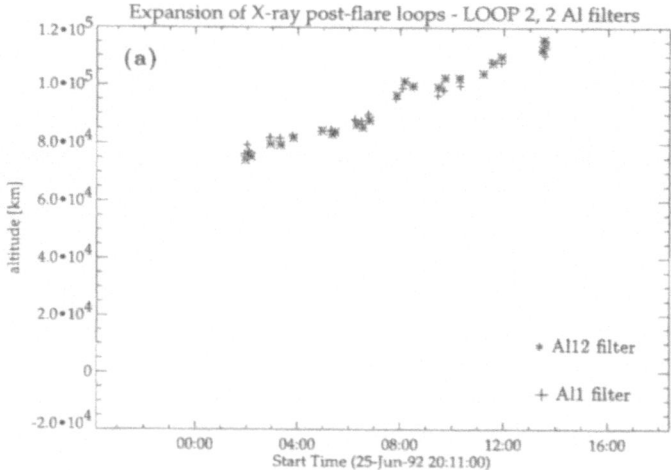

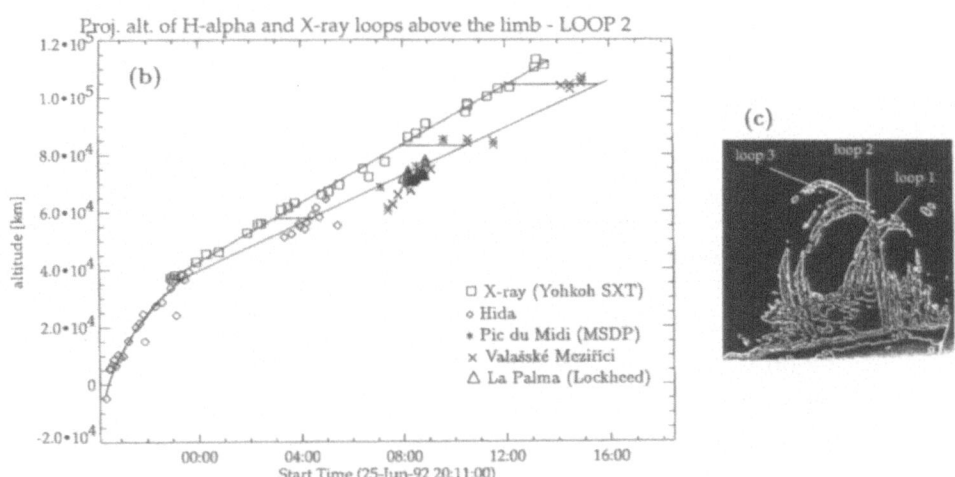

Figure 4. (a) True (reconstructed) altitudes of three main loops observed in X-ray with SXT on June 25-26, 1992 as function of time, (b) Projected altitudes of one of the three post-flare loop systems measured from the limb. In case of no shrinkage of the loops between the X-ray and Hα stages, horizontal lines indicate the observed cooling times. Note that these "observed" cooling times are longer than the calculated ones (cf. Table I). (c) shows the notation of the loops. The error bars of the measurements for Hida and Czech data are between 5 to 10 %, less for the other measurements. Note that the effect of solar rotation has not been taken into account in Figure b. (c) Denotation of the loops.

of the X-ray loops, and computed the cooling times taking into account thermal conduction and radiative losses with a starting temperature of 6.5 $\times 10^6$ K. The computed ("theoretical") cooling times can be compared to "observed" cooling times (Table I), which are obtained from the relative altitude curves of hot and cool loops (Figure 4 b). We see in Figure 4 that at any given time hot loops appear higher than the cool loops, and that the gap between the two altitude curves increases with time. If we suppose that loops do not shrink between the "mature" X-ray loop stage and the Hα stage (which does not exclude the possibility that hot loops shrink right after their formation, as found by Forbes and Acton [6]), then the "observed" cooling time is the difference between the times at which hot and cool loops reach the same altitude. Looking at Table I, it is obvious that there is a good agreement between the computed and observed cooling times 3 and 17 hours after the impulsive phase, while the observed cooling times appear to be longer than the computed cooling times 8 to 12 hours after the onset of the flare. This discrepancy may or may not be real, taking into account the difficulties of such relative altitude measurements using data from 6 quite different sources. Švestka et al., [8] suggested, that since the density must be increasing during the life of a cooling loop, implying that the loop can not stay at the same altitude, but must shrink. Further observations are needed to solve this puzzle.

Table I. Measured and estimated cooling time of hot post-flare loops

Observations			Theory
time	Δt	n_e^{hot} (cm^{-3})	Δt_c
23:00 UT	$\simeq 10 min$	$7\text{-}8^{10}$	10 min
04:00 UT	1 - 2 hours	$2\text{-}3\ 10^{10}$	30 min
08:00 UT	1 - 2 hours	10^{10}	1 hour
13:00 UT	2 hours	$6\text{-}7\ 10^{9}$	2 hours

Hanaoka et al., [9], observed post-flare loops during the February 16, 1980 total eclipse from Kenya. Their Four-Channel Monochromatic Image Telescope was used to obtain FeX 6374, Fe XIV 5303, Hα and λ6100 continuum images, providing observations of the loops in the temperature range of $10^6 - 10^4$ K. Such observations provide an excellent snapshot of the loops at different stages of the cooling process. Since in 1999 solar activity expected to be close to maximum, there is a certain chance that post-flare loops will be present on the Sun, hopefully in the vicinity of the limb. Finer details of the cooling process could then be revealed through joined observations between spacecrafts and ground-based observers of such events, covering a large range in wavelengths (i.e., temperatures), from soft X-rays (Yohkoh) through different UV lines, including the Lyman lines (SOHO/SUMER,

SOHO/CDS) to Hα, CaII (IR) and other optical lines observable from the ground. Observations taken at different places along the totality zone could provide a longer temporal coverage of the event. Such observing programs would greatly increase our understanding of these spectacular events and, among others, could provide an answer to the question if loops shrink or not.

Acknowledgments The authors wish to thank the team of *Yohkoh*/SXT for the SXT data, R. Kitai and Y. Funakoshi for the Hida Observatory data, L. Lenžă and P. Heinzel for the Valašské Meziříci Observatory data, B. Rompolt for Wroclav observations, R.A. Shine and Z. Franck for the La Palma (Lockheed) data and T.Roudier for the Pic du Midi observations. We thank P. Démoulin and T. Forbes for valuable discussions and W. van Driel for critical reading of the paper. LvDG's acknowledges Hungarian research grant T17325 OTKA.

References

1. Švestka, Z. (1989), *Solar Phys.* **121**, 399.

2. Schmieder, B., (1992), in Švestka,Z, Jackson V. and Machado M.E. (eds), *Lecture Notes in Physics*, **399**, 1.

3. Sturrock, P.A., (1968), in K.O. Kiepenheuer (ed.), "Structure and Development of Solar Active Regions", *IAU Symp.* **35**, 471.

4. Kopp, R.A., & Pneuman, G.W. (1976), *Solar Phys.* **50**, 85.

5. Forbes, T.G. & Malherbe, J.M. (1986), *Astrophys. Journ.*, **302**, L67.

6. Forbes, T.G. & Acton, L.W. (1996), *Astrophys. Journ.*, **459**, 330.

7. Moore, R. & 15 co-authors, (1980), in P. Sturrock (ed), "Solar Flares", A Monograph from Skylab Solar Workshop II., p. 341.

8. Švestka, Z., Fontenla,, J.M., Machado, M.E., Martin, S.F., Neidig, D.F., & Poletto, G., (1987), *Solar Phys.* **108**, 237.

9. Hanaoka, Y., Kurokawa, H., & Saito, S. 1986, *Solar Phys.* **105**, 133.

10. Schmieder, B, Heinzel, P., Wiik, J.E., Lemen, J.R., Anwar, B., Kotrč, P. & Hiei, E., (1995), *Solar Phys.* **156**, 337.

11. Schmieder, B., Heinzel, P., van Driel-Gesztelyi, L., & Lemen, J.R., (1996), *Solar Phys.* **165**, 303.

12. van Driel-Gesztelyi, L., Wiik, J.E., Schmieder, B., Kitai, R., Tarbell, T., Anwar, B., Funakoshi, Y.: (1996), *Solar Phys.*, in press

13. Feldman U., & Seely J. F. (1995), *Astrophys. Journ.*, **450**, 90.

14. Wiik, J.E., Schmieder, B., Heinzel, P., & Roudier, T., (1996), *Solar Phys.* **166**, 89.

CHROMOSPHERIC AND CORONAL HEATING MECHANISMS

P. ULMSCHNEIDER
Institut für Theoretische Astrophysik, Universität Heidelberg,
Tiergartenstr. 15, D--69121 Heidelberg, Germany

1. Introduction

The present work discusses the basic physical mechanisms which produce the heating of stellar chromospheres and coronae. As extensive reviews on this subject have been published elsewhere (Narain & Ulmschneider [2], (Paper I), Narain & Ulmschneider [3] (Paper II) as well as Ulmschneider [4] (Paper III)) the present paper concentrates on the physics of the heating processes. For additional literature as well as for additional reviews see Papers I to III. Section 2 discusses the necessity of chromospheric and coronal heating and describes the elementary heating processes. Section 3 outlines the hydrodynamic heating mechanisms and Section 4 the magnetic heating mechanisms.

2. Necessity of Mechanical Heating of Chromospheres and Coronae

Consider a gas element in the chromosphere. The amount of heat dQ *(erg $/cm^3$)* flowing across its boundaries raises the entropy in the element by $dS = dQ/\rho T$, where ρ is the density and T the temperature. For a plane parallel atmosphere one can write the entropy conservation law

$$\rho T\left(\frac{\partial S}{\partial t} + \frac{\partial S}{\partial z}\right) = \Phi_R + \Phi_C + \Phi_V + \Phi_J + \Phi_M \tag{1}$$

Here z is height, t time, v the gas velocity and Φ heating rates *(erg$/cm^3$ s)*. Eq. (1) states that the entropy in the gas element increases if from the outside, energy is supplied by radiative, thermal conductive, viscous, Joule and mechanical heating. Mechanical heating comprises all processes which convert nonradiative, nonconductive hydrodynamic or magnetic energy (henceforth called mechanical energy) flowing through the element into microscopic random thermal motion.

As the chromosphere exists on the Sun for billions of years one can neglect the term $\partial S/\partial t$ in Eq. (1) in a time-averaged model. With a solar wind mass loss

Z. Mouradian and M. Stavinschi (eds.),
Theoretical and Observational Problems Related to Solar Eclipses, 95-102.
©1997 *Kluwer Academic Publishers.*

rate of $\dot{M} = 10^{-14} M_\odot/y$ one can compute a wind speed $v = \dot{M}/(4\pi\rho R^2_\odot) \approx 1.1\ 10^{-11}/\rho$, from which one finds very low flow speeds. Even in the transition layer e.g. with $\rho = 2.3\ 10^{-15} g/cm^3$ and $T=4.5\ 10^5$ K one has a flow speed of only $v=4.7\ 10^3$ compared to a sound speed of $c_S=1\ 10^7$ cm/s. Thus the entire LHS of Eq. (1) can be neglected. Below we will show that in the chromosphere Φ_C, Φ_V and Φ_J can also be neglected, while Φ_C is important in the transition layer and corona. Using a gray expression for the radiative heating rate we thus find for the chromosphere

$$4\pi\overline{\kappa}(\overline{J} - B) + \Phi_M = 0 \qquad (2)$$

Here $\overline{\kappa}$ is the Rosseland opacity, $\overline{J}$ the frequency integrated mean intensity, $B = \sigma T^4/\pi$ the frequency integrated Planck function and σ the Stefan-Boltzmann constant. In the special case of *radiative equilibrium* one has $\Phi_M = 0$ and finds $\overline{J} = B$. On the stellar surface one has $\overline{J} = \sigma T_{eff}^4 / 2\pi$, where T_{eff} is the effective temperature and the factor 1/2, as there is intensity only away from the star. In radiative equilibrium one therefore finds $T= 1/2^{1/4}\ T_{eff} \approx 0.8\ T_{eff}$, that is, in absence of mechanical heating the outer stellar regions would have temperatures of the order of the boundary temperature. However, as a chromosphere is a layer where the temperature rises in outward direction to values $T >> T_{eff}$, it is clear that $B >> \overline{J}$ and therefore one must have $\Phi_M >> 0$. This shows that for chromospheres mechanical heating is essential. In addition, as the energy loss of the transition layer and corona cannot be balanced by thermal conduction from a reservoir at infinity, but must ultimately be supplied from the stellar interior, we conclude that for transition layers and coronae, mechanical heating is also essential.

Moreover, chromospheres and coronae can only be maintained if mechanical heating is constantly applied. The time scale, in which an excess temperature will cool down to the boundary temperature, if the mechanical heating were suddenly disrupted, is given by the *radiative relaxation time*

$$t_{Rad} = \frac{\Delta E}{\Phi_R} = \frac{\rho c_v \Delta T}{16\overline{\kappa}T^3\Delta T} = \frac{\rho c_v}{16\overline{\kappa}\sigma T^3} \approx 1.1\cdot 10^3\ s \qquad (3)$$

Here from a typical solar model at $z = 1280$ km: $T= 6200$ K, $p = 4.4\ dyn/cm^2$, $\overline{\kappa}/\rho = 4.1\ 10^{-4}\ cm^2/g$, $c_v = 9.6\ 10^7\ erg/gK$, $\sigma =5.6\ 10^{-5}\ erg /cm^2\ s\ K^4$. It is seen that in timescales of a fraction of an hour the chromosphere would cool down to the boundary temperature if mechanical heating would suddenly be interrupted.

Table1 summarizes the mechanisms which are thought to provide a steady supply of mechanical energy to balance the chromospheric and coronal losses. Here occasional transient and localized heating events like large flares are not considered because they do not contribute appreciably to the persistent chromospheric and coronal heating. The term heating mechanism comprises three physical aspects, the *generation* of a carrier of mechanical energy, the *transport* of mechanical energy into the chromosphere and corona and the *dissipation* of this energy in these layers. Table.1 shows the various proposed energy carriers which can be classified into two main categories as *hydrodynamic* and *magnetic* mechanisms. The magnetic mechanisms can be subdivided further into wave- or AC-mechanisms and current sheet- or DC-mechanisms. Also in Table.1 the mode of dissipation of these mechanical energy carriers is indicated. Ultimately these mechanical energy carriers derive their energy

TABLE 1 Mechanical heating mechanisms for stellar chromospheres and coronae, P is the wave period and P_A the acoustic cut-off period.

energy carrier	dissipation mechanism
hydrodynamic heating mechanisms	
acoustic waves, $P < P_A$	shock dissipation
pulsational waves, $P > P_A$	shock dissipation
magnetic heating mechanisms	
1. alternating current (AC) or wave mechanisms	
slow mode mhd waves,longitudinal mhd tube waves	shock dissipation
fast mode mhd waves	Landau damping
Alfvén waves (transverse, torsional)	resonance heating
	compressional viscous heating
	turbulent heating
	Landau damping
magnetoacoustic surface waves	mode-coupling
	phase-mixing
	resonant absorption
2. direct current (DC) mechanisms	
current sheets	reconnection
	(turbulent heating, wave heating)

from the nuclear processes in the stellar core from where the energy is transported in he form of radiation and convection to the stellar surface. In late-type stars the mechanical energy generation arises from the gas motions of the surface convection zones. These gas motions are largest in the regions of smallest density near the top boundary of the convection zone. Due to this the mechanical energy carriers, particularly the waves, are generated in a narrow surface layer.

Let us now discuss the *elementary heating processes*. In the dissipation process, mechanical energy is converted into heat. That is, organized motion or potential energy is converted into random thermal motion. As will be shown below, an efficient conversion process is almost always associated with the generation of large variations of the physical variables over very small scales. For instance, it has been known for a long time that an efficient way to dissipate acoustic waves is the formation of shocks, where the physical variables abruptly vary over distances of a molecular mean free path.

Consider a typical acoustic or magnetohydrodynamic disturbance in the solar chromosphere with characteristic parameters, size $\Delta L = 200$ km, temperature $\Delta T = 1000K$, velocity $\Delta v = 3$ km/s and magnetic field perturbation $\Delta B = 10$ G. Using appropriate values for the thermal conductivity $\kappa_{T\eta} = 10^5$ erg/cm s K, viscosity $\eta_{vis} = 5 \, 10^4$ dyn s/cm^2 and electrical conductivity $\lambda_{el} = 2 \, 10^{10}$ s^{-1} we find for the thermal conductive heating rate

$$\Phi_C = \frac{d}{dz} \kappa_{th} \frac{dT}{dz} \approx \frac{\kappa_{th} \Delta T}{\Delta L^2} \approx 3 \cdot 10^{-7} \quad \left[\frac{erg}{cm^3 s} \right] \tag{4}$$

the viscous heating rate

$$\Phi_V = \eta_{vis} \left(\frac{dv}{dz} \right)^2 \approx \frac{\eta_{vis} \Delta v^2}{\Delta L^2} \approx 1 \cdot 10^{-7} \quad \left[\frac{erg}{cm^3 s} \right] \tag{5}$$

the Joule heating rate:

$$\Phi_J = \frac{J^2}{\lambda_{ei}} = \frac{c_L^2}{16\pi^2 \lambda_{el}} \left(\nabla \times B \right)^2 \approx \frac{c_L^2 \Delta B^2}{16\pi^2 \lambda_{el} \Delta L^2} \approx 7 \cdot 10^{-5} \left[\frac{erg}{cm^3 s} \right] \tag{6}$$

Here J is the current density and c_L the light velocity. The three heating rates show that normally these processes are inadequate to balance the empirical chromospheric cooling rate of $-\Phi_R = 10^{-1}$ erg/ cm^3 s. Only when the length scale ΔL is considerably decreased, can the heating rates be raised to acceptable levels. For acoustic waves as well as slow mode mhd- and longitudinal mhd tube waves, this is accomplished by shock formation. For magnetic cases, by the formation of current sheets.

3. Hydrodynamic Heating Mechanisms

There are two hydrodynamic mechanisms acoustic waves and pulsational waves. Acoustic waves have periods less then the acoustic cut-off period $P_A = 4\pi \, c_S / (\gamma \, g)$ while pulsational waves have periods $P > P_A$ (see Papers I and II for details). Acoustic waves are generated in the surface convection zones of late-type stars and by radiative instabilities in early-type stars, they heat by shock dissipation. Pulsational waves, most prominent in Mira-star pulsations, but also in other late-type giants, are generated by the *kappa-mechanism* and related processes. Like acoustic waves, pulsational waves heat by shock dissipation.

4. Magnetic Heating Mechanisms

4.1. MODE-COUPLING

This mechanism is not a heating process by itself, but converts wave modes, which are difficult to dissipate, by non-linear coupling into other modes, where the dissipation is more readily achieved. Typical cases are the conversion of transverse or torsional Alfvén waves into acoustic-like longitudinal tube waves. For examples of these two processes see Papers I and II.

4.2. RESONANCE HEATING

Resonance heating occurs, when upon reflection of Alfvén waves at the two foot points of the coronal loops, one has constructive interference. For a given loop length $l_{||}$ and Alfvén speed c_A, resonance occurs, when the wave period is $mP = 2\, l_{//}\, c_A$, m being a positive integer. Waves which fulfill the resonance condition are trapped and after many reflections are dissipated by Joule-; thermal conductive or viscous heating. Examples of this process are given in Papers I and II.

4.3. COMPRESSIONAL VISCOUS HEATING

Compressional viscous heating, recently proposed by Strauss [7] (see also Paper II), is a very promising mechanism for coronal regions, where the gyro frequency is much larger than the collision frequency. Swaying an axial magnetic flux tube sideways with velocity v results in a transverse Alfvén wave which is incompressible $\nabla \cdot v = 0$ to first order. This is different for tubes with helicity, where one has $\nabla \cdot v \approx \rho\,/\,\rho$. With an increase of the density, the magnetic field is compressed and the gyro frequency increased. Gyrating around the field lines more quickly, the ions after colliding with each other, generate larger velocities in non-perpendicular directions as well, which constitutes the heating process.

4.4. TURBULENT HEATING

In a turbulent flow field with high Reynolds number there are bubbles of all sizes. The energy usually is put into the largest bubbles. Because of the large inertial forces the big bubbles are ripped apart into smaller bubbles, and these in turn into still smaller ones etc. This process is called turbulent cascade. A turbulent flow field can be described by three characteristic quantities, density ρ_0, bubble scale $l_k = 2\pi/k$, and the mean velocity u_k of such bubbles. k is the wavenumber. It is easily seen, that from these three quantities only one combination for a heating rate can be formed:

$$\Phi_k = \rho_0 \frac{u_k^3}{l_k} \quad \left[\frac{erg}{cm^3 s} \right] \tag{7}$$

If there are no other losses, like by radiation, all the energy which is put in at the largest bubbles must reappear in the smaller bubbles etc. Thus if $k1$, $k2$, ... represents a series of smaller and smaller bubbles one must have $\Phi_{\kappa1} = \Phi_{\kappa2} = ... =$ const. This implies

$$u_k \sim l_k^{1/3} \tag{8}$$

which is the *Kolmogorov law*. The range $l_{k1}...l_{k2}$ of validity of this law is called the *inertial range*. Consider what happens if l_k becomes very small. From Eqs. (8) and (11) one finds for the viscous heating rate $\Phi_V = \eta_{vis} (du / dl)^2 \approx \eta_{vis} u_k^2 / l_k^2 \approx \eta_{vis} l_k^{-4/3}$ which goes to infinity for $l_k \rightarrow 0$. Thus at some small enough scale, viscous heating sets in and the inertial range ends. It is seen that turbulent heating lives from the formation of small scales. One can visualize the process as follows. Because of the continuous splitting of bubbles into smaller sizes, with the velocities decreasing much less rapidly, one eventually has close encounters of very small bubbles with large velocity differences where viscous heating dominates. In magnetic flux tubes the turbulent cascade is supposed to happen in transverse direction to the field (see Papers I and II).

4.5. LANDAU DAMPING

Landau damping occurs at coronal heights, where the collision rate becomes small. As Chen ([1] Fig. 7.17) has well explained, this process is analogous to surfing on ocean waves. When surfing, a surfboard rider launches himself in propagation direction into the steepening part of an incoming wave and gets further accelerated by this wave. In Landau damping, the propagating wave accelerates gas particles which, due to their particle distribution function, happen to have similar direction and speed as the wave. Because a distribution function normally has many more slower particles than faster ones, the wave looses energy to accelerate the slower particles. This gained energy is then shared with other particles in the process to reestablish the distribution function which constitutes the heating mechanism.

4.6. RESONANT ABSORPTION

In the process of resonant absorption one considers magnetoacoustic surface waves in a magnetic field B which points in z-direction, and varies from B_1 to B_2 in x-direction (see Figure 1). The surface wave, with its field perturbation $\delta B = B'_x$ in x-direction, has a phase speed $\upsilon_{ph} = ((B_1^2 + B_2^2) / (4\pi(\rho_1 + \rho_2)))^{1/2}$, such that at an intermediate position x_0, the phase speed becomes equal to the local Alfvén speed

$c_{Ao} = B(x_0) / \sqrt{4\pi\rho(x_0)}$. In panel a of Figure -1 consider the wave fronts of the peak (drawn) and trough (dotted) of a surface wave. Because to the right of x_0, the Alfvén speed is larger and to the left smaller, the wave fronts at a later time get tilted, relative to the phase, propagating with speed c_{Ao} (see panel b). At a still later time (panel b) the wave fronts get tilted even further and approach each other closely at the position x_0. This leads to small scales and intense heating at that field line. For the recent extensive analytical and numerical work on this heating process see Papers I and II as well as the references therein.

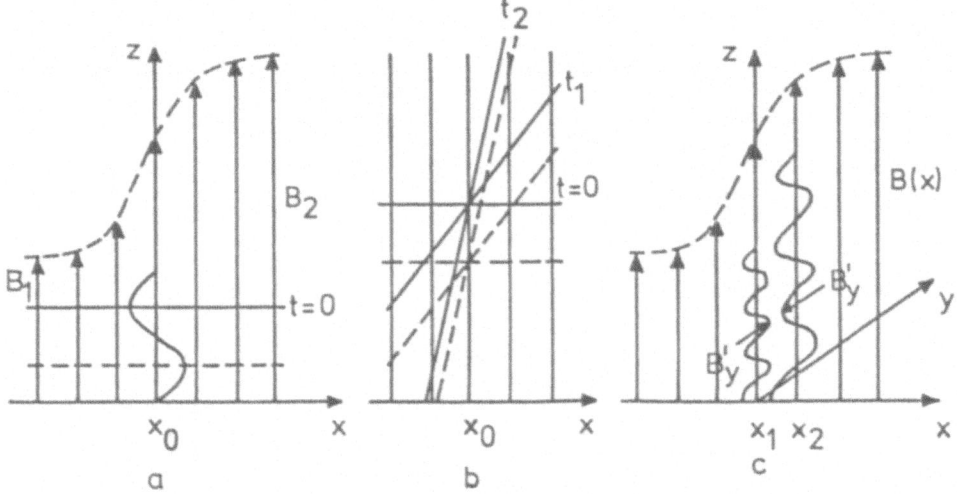

Figure 1 In a field pointing in z-direction, where the field strength varies in x-direction: a) resonant absorption of a surface wave (shaking in x-direction), wave fronts at time t=0, b) these wave fronts at subsequently later times t_1, and t_2, c) phase-mixing of a surface wave (shaking in y-direction).

4.7. PHASE-MIXING

For phase-mixing (c.f. panel c of Figure 1) one considers the same magnetic field geometry as in panel a of Figure 1, however, the field perturbation $\delta B=B'_x$ of the wave is now in y-direction, perpendicular to the x- and z-directions. As the Alfén speeds of two closely adjacent regions x_1 and x_2 in x-direction are different, it is seen that after propagating some distance Δz, the fields $B'_y(x_1)$ and $B'_y(x_2)$ will be very different, leading to a current sheet and strong dissipation. Here again it is the appearance of small scale structures which lead to dissipation (see Papers I and II).

4.8. RECONNECTION

As examples of the DC heating mechanisms I discuss two situations where current sheets are thought to exist [6], shows an arcade system, which by slow motion is laterally compressed and develops a current sheet, where oppositely directed fields reconnect. The other example by Parker [5] pictures a tangled and braided web of coronal loops created by slow foot point motions. As the motions put in more and more energy, the system tries to return to its minimum energy configuration. This can only be done by reconnection. At many locations in the web, oppositely directed fields occur giving rise to local current sheets, which by reconnection (in the form of microflares) release the magnetic field energy. The energy is dissipated both directly and via the generation of waves and turbulence. Note that reconnection likewise happens in small scale regions.

References

1. Chen, F.F. (1984), *Introduction to Plasma Physics and Controlled Fusion* 2^{ed}, Vol 1, Plasma Physics,Plenum Press, New York
2. Narain, U., Ulmschneider, P. (1990), *Space Sci. Rev.* **54**, 377 (Paper I)
3. Narain, U., Ulmschneider, P.); (1996), *Space Sci. Rev.* **75**, 453 (Paper II)
4. Ulmschneider, P. (1996), in: *Cool Stars Stellar Systems and the Sun*, ASP Conf. Ser., R. Pallavicini, A.K. Dupree (eds.) in press (Paper III)
5. Parker, E.N. (1992), *J. Geophys. Res.* **97**, 4311
6. Priest, E.R. (1991), in *Mechanisms of Chromospheric and Coronal Heating*, Ulmschneider P., Priest E.R., Rosner R., (eds.) Springer, Berlin, p 520
7. Strauss, H.R. (1991), *Geophys. Res. Let.* **18**, 77

RECONSTRUCTION OF THE LARGE-SCALE DISTRIBUTION OF CORONAL ELECTRONS FROM ECLIPSE DATA

F. CLETTE
Observatoire Royal de Belgique,
Brussels, Belgium

Abstract. Mathematical models of the global electron density distribution in the corona were first constructed from solar eclipse images at the end of the last century. Since then, the complexity of these density models has increased steadily, as additional free parameters and new mathematical tools were incorporated. The ultimate goal of this effort has always been to improve the representation of the inhomogeneous coronal structure, while maintaining a restricted set of parameters. This review puts the successive steps of this maturation process in a general perspective. A recent model, developed at the Royal Observatory of Belgium for the 1991 and 1994 eclipses, is described to illustrate the modeling techniques and some current issues.

1. Introduction

The analysis of global properties of the solar corona, using polarized white-light observations made during solar eclipses, began about one hundred years ago. Since the true nature of the diffusing particles, i.e. the free electrons, was discovered, much progress has been achieved in our understanding the overall density distribution of the coronal plasma and of its time evolution. This long-term effort is justified by the multiple repercussions of large-scale models of the coronal electron density in the field of solar physics. These models help in the interpretation of coronal observations at radio and X-ultraviolet wavelengths, and they also provide constraints on the plasma density, temperature and velocity [1, 2, 3]. Furthermore, these global distributions are representative of the concept of the so-called "background corona" [4] and constitute the first step in the determination of the F corona distribution.

During the last decade, the availability of new computing tools brought a new impulse to this kind of research, and new models were recently introduced with a degree of complexity that allows a more detailed and more realistic representation of the inhomogeneous coronal medium. This evolution will be briefly retraced here.

2. Fundamental Integral Equations and the "Power Law" Family of Solutions

The tangential and radial components of the observed K corona intensity, K_t and K_r, are given by the following integral equations:

Z. Mouradian and M. Stavinschi (eds.),
Theoretical and Observational Problems Related to Solar Eclipses, 103-109.
©1997 *Kluwer Academic Publishers.*

$$K_t(x) = 2\pi\sigma\overline{B_\Theta}R_\Theta\int_x^\infty N_e A(r)\frac{r\,dr}{\sqrt{r^2-x^2}}$$

$$(K_t - K_r)(x) = 2\pi\sigma\overline{B_\Theta}R_\Theta\int_x^\infty N_e[A(r)-B(r)]\frac{x^2\,dr}{r\sqrt{r^2-x^2}} \tag{1}$$

where r is the radial distance, x is the elongation from disc center, σ is the Thomson diffusion coefficient, $\overline{B_\Theta}$ is the average solar-disk intensity and R_Θ is the solar radius. Those fundamental integral equations, where $A(r)$ and $B(r)$ are the dilution functions defined by Van de Hulst [5], express the integration of the diffused intensity along the line of sight.

The mathematical framework of this problem was established in 1879 by Schuster [6], who introduced an electron distribution with the following radial dependency:

$$N_e(r) = \sum_{i=1}^{I} \frac{a_i}{r^{b_i}} \tag{2}$$

where the coefficients a_i and the exponents b_i are the parameters that must be determined, with I typically equal to 3.

This idea had important consequences, as almost all models developed since then use this power law dependency, even though it has no firm physical basis. However, this choice is justified because an analytical solution to the integrals can be found. Indeed, with only a limited number of initial assumptions and simplifications, the integrals in Equ.1 can be split into elementary integrals: elliptic integrals with integer b exponents [6] or series expansions in terms of the S_n integrals defined by Saito [7].

Fifty years after the foundation work of Schuster, Minnaert [8] proposed the first physical model of the corona, by incorporating the theory of Thomson scattering by free electrons. Like Schuster, Minnaert assumes integer powers and uses elliptic integrals. Baumbach [9] and Van de Hulst [5] both start by representing the observed intensity distribution, instead of the final density distribution, by three $1/r^b$ terms. They then solve the Abel integral equation to obtain an expression of the local diffused intensity, which takes the same mathematical form (three radial terms) but with different coefficients containing the contribution of the dilution functions. To get the final electron distribution, Baumbach uses an approximation: he develops the dilution functions in Taylor series and keeps only five terms. Van de Hulst defines an iterative technique to derive rigorously the optimal function, i.e. the solution providing a coherent representation of the intensity and polarisation simultaneously.

All the above distributions assumed a spherical symmetry (Figure 1.a), which soon proved to be inadequate for the proper representation of the flattened corona observed at the minima of the solar activity cycle. Saito [7] introduced the first non-spherical model to improve the fit to the minimum-type coronae (Figure 1.b). His axisymmetrical model again assumes the same kind of radial variations and adds a simple latitude dependency to each radial term, as follows:

$$N_e(r,\varphi) = \sum_{i=0}^{3} N_{0,i} \frac{1-f_i \sin^{s_i} \varphi}{r^{b_i}} \tag{3}$$

where the factors f_i and the exponents s_i are new parameters.

Starting from this density distribution, he deduces a general expression for the observables K_t and K_t - K_r a series expansion of the elementary S_n integrals defined in his paper.

It is essential to note here, that, most of these authors adjust the model parameters manually by a trial and error method, in order to fit their models to the data. In this respect, Van de Hulst is first to propose a self-consistent iterative scheme. In the last 20 years, only one additional model was built according to the above scheme by Perry and Altschuler [10,11]. As they were analyzing coronagraph data that contain additional information about the longitude variations of the density, the angular distribution is fully developed in terms of spherical harmonics. This model is solved by a least-square method and leads to the most sophisticated 3-dimensional mapping of the low corona: the original model was based on 128 parameters, while the revised 1989 model depends on 23000 parameters.

3. The SGCC Model

In the above latitude-dependent models, the symmetry axis of the density distribution is always supposed to coincide with the rotation axis of the Sun and effects of the small inclination ($< 7°$) are simply neglected. However, as pointed out by Gulyaev and Vanyarkha [12] and Sime and McCabe [13], the highest coronal densities are actually observed along the neutral sheet marking the equator of the global dipolar magnetic field of the Sun (streamer belt). Furthermore, during most of the solar cycle, the magnetic poles are strongly inclined relative to the solar rotation axis, by up to 60°. Clearly, the arbitrary choice of the rotation axis does not reflect the physical structure of the global corona and high inclinations must be incorporated explicitly in the model calculations.

With this goal in mind, a new latitude-dependent model was developed by J.-P. Scandariato, J.-R. Gabryl, F. Clette and P. Cugnon (SGCC) [14]. It is based on the following expression:

$$N_e(r,\varphi) = \sum_{n=0}^{N_{max}} P_n(\sin\varphi) \sum_{i=0}^{I_{max}(n)} \frac{a_{i,n}}{r^{b_{i,n}}} \tag{4}$$

106

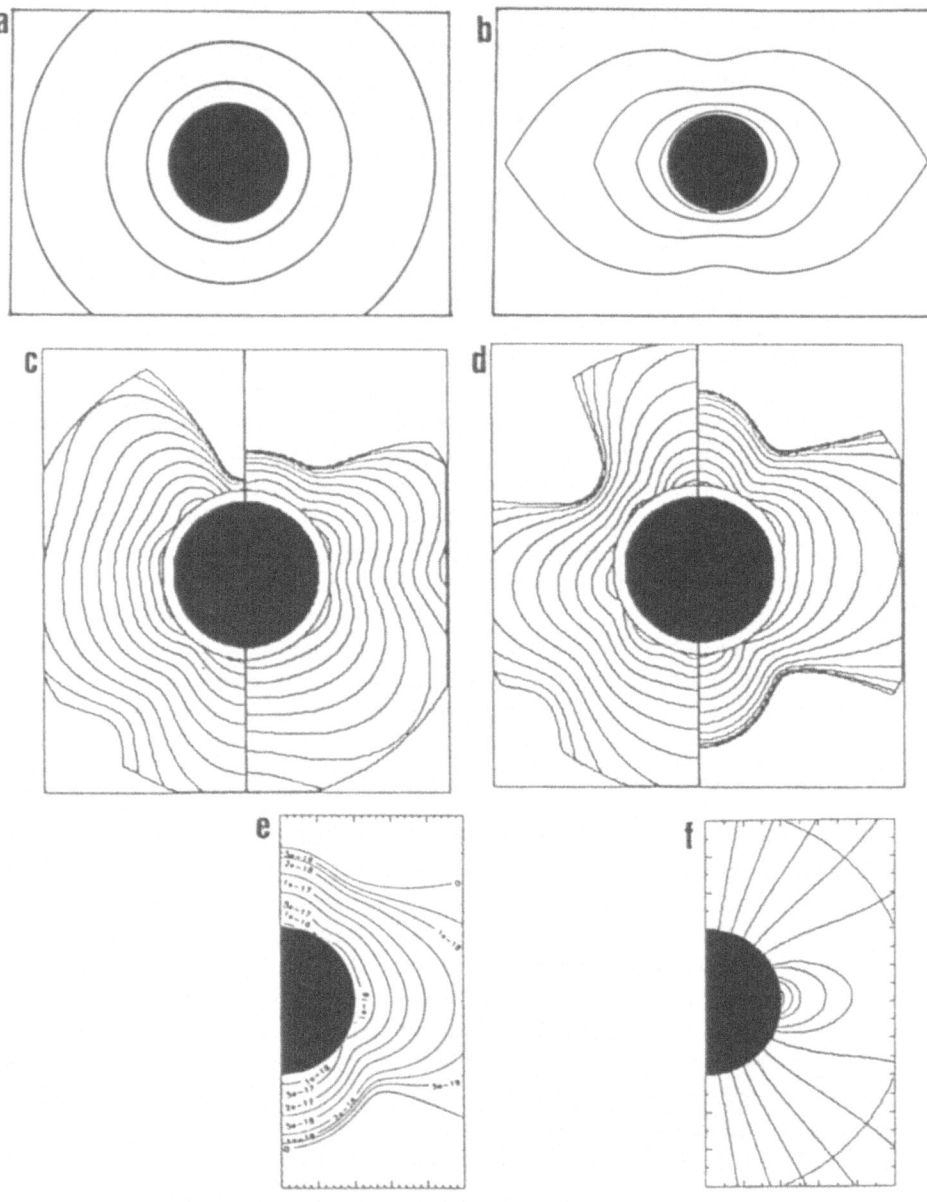

Figure1. The evolution of global models of the electron density (isodensity maps): early spherical distributions (a), Saito model (b), SGCC model for the 1991 eclipse (c: aligned on the rotation axis, d: aligned on the dipolar magnetic field), Gibson-Bagenal model (e: electron density, f: magnetic field).

where the latitude variations are projected on the orthogonal Legendre polynomials P_n. Using this fairly general representation, with a separate radial function for each angular term, it becomes possible to reproduce more complicated latitude variations than a simple polar flattening, including North-South asymmetries. At the same time, the unrealistic discontinuities produced by Saito's formulation at the solar equator are avoided.

In order to allow for any inclination of the symmetry axis of the distribution relative to the line of sight, the rotation of the coordinate system is treated explicitly in the integral equations. By applying the addition theorem for Legendre polynomials and by developing the $A(r)$ and $B(r)$ dilution functions in Taylor series (35 terms), the integrals again reduce to a sum of elementary integrals, which can be solved in terms of Gamma functions. This solution presents an interesting property: by rearranging the order of summations, the radial and angular terms can be separated. This leads to this final expression:

$$K_t(r,\psi) = \sum_{n=0}^{N_{max}} A_m(r)\cos m\psi$$

$$(K_t - K_r)(r,\psi) = \sum_{n=0}^{N_{max}} B_m(r)\cos m\psi$$

(5)

where ψ is the polar angle, m is the angular order, and the $A_m(r)$ and $B_m(r)$ functions gather all purely radial dependencies.

Thanks to the above property, the resolution can be split in two successive steps, using a least square method. The final model contains up to 10 angular orders, with up to 3 radial terms for each, i.e. a maximum of 30 parameters. Figures 1.c/1.d illustrates the results obtained for the 1991 eclipse. Note the increased symmetry of the model aligned on the true magnetic configuration. A more detailed description of those results is provided by Gabryl et al. [15].

4. Numerical and Physical Models

New directions have been explored more recently in order to avoid some limitations of the models described thus far, which require an analytical solution of the integral equations. Due to the high computational requirements, only a few attempts have been made to integrate numerically the expressions in Equ.1. Up to now, this approach was only applied to individual structures, like single streamers [16].

Other authors rightly questioned the absence of direct physical meaning of Schuster's radial expression and tried to introduce a more physical representation of the electron density [17, 18, 19, 20]. Most of these physical models use the hydrostatic approximation, which is valid over most of the intermediate corona (1.2 to 3 $R_\odot$). In this case, the density distribution decreases exponentially as [17]:

$$N_e(r) = N_e(r_\odot) \exp\left[-\frac{\mu m_H g_\odot R_{\odot\odot}^2}{\kappa T} \left(\frac{1}{r_\odot} - \frac{1}{r} \right) \right] \qquad (6)$$

However, with the above definition, an analytical solution can only be found for the integral equations if very crude approximations are made: spherical symmetry, isotropic diffusion coefficient or the replacement of the radial distance r by the elongation x in the dilution functions. The line-of-sight integrals then reduce to Modified Bessel functions [17]. Otherwise, accurate models imply a fully numerical treatment. As a consequence, up to now, such models were only applied to limited and homogeneous regions in the corona, like streamers and polar holes.

More recently, Gibson and Bagenal [21, 22] proposed a new global physical model based on the magnetostatic model of Bogdan and Low [23], which includes explicitly for the first time the magnetic field. They obtain an expression of the electron density distribution in terms of spherical harmonics, with again a $1/r$ radial dependency. This model allowed the reconstruction of a very simple and symmetrical configuration (minimum type corona, no inclination; Figure 1.e/1.f). However, it proved impossible to obtain convergence towards a solution when the photospheric field was applied as a boundary condition.

5. Conclusion

The classical electron density models have reached a high degree of sophistication. Even though their mathematical formulation does not reflect directly the underlying physics, they provide good constraints for studies of the heating mechanisms in the corona and of the solar wind. Presently, they are the only ones that offer enough flexibility to adapt to the highly variable coronal structure and reflect the true symmetry imposed be the heliospheric current sheet. On the other hand, by using hydrostatic and magnetostatic models describing the physical state of the coronal plasma, values for the temperature, pressure and magnetic field can be determined more directly, despite some approximations. However, such models proved to be less flexible and until now, they were applied only to very simple configurations.

In the future, both modeling techniques could be associated, the mathematical model distributions serving for instance as boundary conditions for physical models. Finally, as computing power increases steadily, pure numerical models might offer the best prospects, as no initial assumption on the form of the solution is required. The coronal could then be realistically reconstructed by the combination of any number of discrete structures, with either closed or open field configurations, and superimposed on a smooth background corona similar to the global distributions established by the classical methods.

References

1. Munro, R.H., Jackson, B.V. (1977) Physical properties of a polar coronal hole from 2 to 5 $R_\odot$, *Astrophys. J.* **213**, 874-886.
2. Feldman, W.C. (1976) Implication of Saito's coronal density model on the polar solar wind and heavy ion abundances, *Journal Geoph. Res.* **82**, 667-673.
3. Withbroe, G.L. (1988) The temperature structure, mass, and energy flow in the corona and inner solar wind, *Astrophys. J.* **325**, 442-467.
4. Saito, K., Poland, A.I., Munro, R.H. (1977) A study of the background corona near solar minimum, *Sol. Phys.* **55**, 121-134.
5. van de Hulst, H.C. (1950) The electron density of the solar corona, *Bull. Astron. Inst. Neth.* **XI, 410**, 135-150.
6. Schuster, A. (1879) On the polarisation of the solar corona, *Monthly Notices of the R.A.S.* **40**, 35-57.
7. Saito, K. (1970) A non-spherical axisymmetric model of the solar K corona of the minimum type, *Ann. Tokyo Astron. Obs., Second Ser.* **12**, 53-120.
8. Minnaert, M. (1930) On the continuous spectrum of the corona and its polarisation, *Zeitschrift fur Astrophys.* **1**, 209-236.
9. Baumbach, S. (1939) Die polarisation der sonnenkorona, *Astron. Nachrichten* **B.267, 6401-02**, 273-296.
10. Perry, R.M., Altschuler, M.D. (1973) Improved three-dimensional mapping of the electron density distribution of the solar corona, *Sol. Phys.* **28**, 435-456.
11. Perry, R.M., Altschuler, M.D. (1989) A fast technique for improved 3-dimensional mapping of the solar corona, *Sol. Phys.* **119**, 301-321.
12. Gulyaev, R.A., Vanyarkha, N.Ya. (1992) Regularities of variation of the heliospheric current sheet orientation during the solar activity cycle, *Sol. Phys.* **140**, 369-378.
13. Sime, D.G., Mc Cabe, M.K. (1990) The structure of the white-light corona and the large-scale solar magnetic field, *Sol. Phys.* **126**, 267-284.
14. Gabryl, J.-R. (1994) *Développement d'un modèle de densité électronique de la couronne solaire*, Mémoire, Institut d'Astrophysique, Université de Liège.
15. Gabryl, J.-R., Cugnon, P., Clette, F. (1996) Comparing the large-scale coronal electron density distribution of the 1991 and 1994 solar eclipses, *these proceedings*.
16. Bohlin, J.D., Garrison, L.M. (1974) Numerical calculation of Thomson scattering from inhomogeneous models of the corona, and application to streamers of the 1970 and 1972 eclipses, *Sol. Phys.* **38**, 165-179.
17. Badalyan, O.G., Livshits, M.A. (1985) White emission of the hydrostatic corona, *Sov. Astron* **29(1)**, 76-81 (*Astron. Zh.* **62**, 132-140).
18. Badalyan, O.G., Livshits, M.A. (1986) The K-corona under hydrostatic density distribution: relevance to the solar wind, *Sol. Phys.* **103**, 385-392.
19. Dzifcakova, E., Rusin, V., Rybansky, M. (1986) The solar corona during the solar eclipse of February 16, 1980. (Density, temperature and expansion velocity), *Bull. Astron. Inst. Czech.*, **37**, n°5, 253-260.
20. Guhathakurta, M., Rottman, G.J., Fisher, R.R., Orrall, F.Q., Altrock, R.C. (1992) Coronal density and temperature structure from coordinated observations associated with the total solar eclipse of 1988 March 18, *Astrophys. J.* **388**, n°2, 633-643.
21. Bagenal, F., Gibson, S. (1991) Modeling the large-scale structure of the solar corona, *Journal Geoph. Res.* **96**, n° **A10**, 17663-17674.
22. Gibson, S., Bagenal, F. (1994) Large-scale coronal magnetic field and density structures, in Rusin, V., Heinzel, P. and Vial, J.-C. (eds.), *Proc. IAU Colloq.* **144**, "Solar coronal structures", VEDA Publ., 155-157.
23. Bogdan, T.J., Low, B.C. (1986) Three-dimensional structures of magnetostatic atmospheres. II. Modeling the large-scale corona, *Astrophys. J.* **306**, 271-283.

A NEW UNDERSTANDING OF THE CORONAL SHAPE CHANGES DURING THE SOLAR CYCLE

J. SYKORA
Astronomical Institute, Slovak Academy of Sciences
05960 Tatranska Lomnica, Slovak Republic

P. AMBROZ
Astronomical Institute, Czech Academy of Sciences
25165 Ondrejov, Czech Republic

1. Introduction

Shape and flattening of the solar corona observed on July 11, 1991 eclipse and confrontation of them with calculated topology of coronal magnetic field and with known variations of the coronal flattening index as a function of the phase of sunspot cycle initiated doubts on acceptability of Ludendorff's definition of the solar corona flattening [7] and, consequently,on real extent of the mentioned variations [9,3]. The following points are crucial for a new approach to the physical understanding of the coronal shape changes throughout the solar cycles:

   - Evolution of the coronal and heliospheric magnetic field topology within solar cycles can be hopefully derived from distribution of magnetic field measured at the level of solar photosphere. Hoeksema [5] shows very clearly that, typically, the neutral magnetic line (better to say, the neutral magnetic plane) is flat and almost identical with the solar equator near to the solar cycle minima, while, the same plane is far more goffered and highly inclined to the equator in the periods around the solar cycle maxima.

   - Helmet-like streamers represent the brigtest and most pronounced features, pre-determining almost solely the shape of the solar corona above $2R_\odot$ (see, e.g., Figure 3 in Koutchmy et al.,[6]). At the same time, they are inherently magnetic structures [8] creating more or less continuous belt around the sun and situated just above the mentioned neutral magnetic plane. Understandably, space orientation of this belt of streamers strictly follows evolution of the global magnetic field described by Hoeksema [5].

   - It is important to realize that the coronal form observed at any solar eclipse represents an apparent two-dimensional projection of the streamers onto the plane of the sky, instead of a real three-dimensional distribution of the streamers in space. Therefore, the observed and real forms of the solar corona may differ strongly,

111

Z. Mouradian and M. Stavinschi (eds.),
Theoretical and Observational Problems Related to Solar Eclipses, 111-115.
©1997 *Kluwer Academic Publishers.*

especially in the periods around the solar cycle maxima. Understandably, degree of this difference at a given day depends mainly from angular distance of the highly inclined magnetic equator from the solar central meridian.

    - A sinusoid-like curve of variations of the coronal flattening index (calculated by using the Ludendorff's definition of this flattening) within the solar cycle was currently presented in the past (see, e.g., Figure 5 in Koutchmy et al.,[6]). Collection of all the coronal flattenings observed during eclipses from 1851 till now provides us with a concept of the flat solar corona at the cycles minima (ellipticity reaches 0.3 at $R_\odot$ from the solar limb), the apparently circular corona at the cycles maxima (the ellipticity being close to zero), and it displays the intermediate forms outside this two extreme periods of the solar cycle curve.

2. Analysis of Coronal Forms Using Eclipse Pictures and Calculated Magnetic Fields of the Solar Corona

We obseved successfully the white-light corona during 1973, 1980, 1981, 1983, 1991, 1994 and 1995 solar eclipses, i.e., one full 22-year Hale magnetic cycle is covered by these observations. Stanford daily measurements (presently Wilcox Solar Observatory) of the photospheric magnetic field (accessible from 1976) enabled us to extrapolate the photospheric field into the solar corona. Description of coronal structures and forms, method of calculation together with discussion of coronal magnetic fields, estimation of coronal flattenings and their confrontation with the really observed shapes of the solar corona were successively published in Sykora and Badalyan,[9]; Ambroz and Sykora, [1,2]; Sykora et al.,[11,12]. The principal results are as follows:

    - Surprisingly good agreement between observed shapes of the corona and calculated coronal magnetic field structures was found, confirming that the streamers are strongly magnetic structures [2,11] .

    -- We have found the Ludendorff's definition of coronal flattening to give very contradictory results in some statistically rare cases (for example, in the periods around the solar cycle maxima and contemporary position of the solar magnetic equator close to the solar meridian). An excellent example of such situation was seen on the July 11, 1991 eclipse day (see the second panel from above in Figure 1 and some other figures in our above-mentioned papers). The solar corona was strongly flattened (however, towards the higly inclined magnetic equator), while, calculation of coronal flattening according to the Ludendorff's definition marks a perfectly circular corona [10,11].

    - We have shown that substantial changes of the apparent shape of the corona have place due to solar rotation. For example, only 3-4 days apart the 1991 eclipse day would be sufficient for this corona to appear circularly symmetrical. This is, of course, due to different projection of the streamers onto the plane of the sky [10]. An opposite case to that of 1991 became on the February 16, 1980 eclipse (upper panel in Figure 1), when solar magnetic equator was seen not from the edge, as in 1991, but more from the face. Thus, the 1980 solar corona unavoidably appeared considerably circular [10].

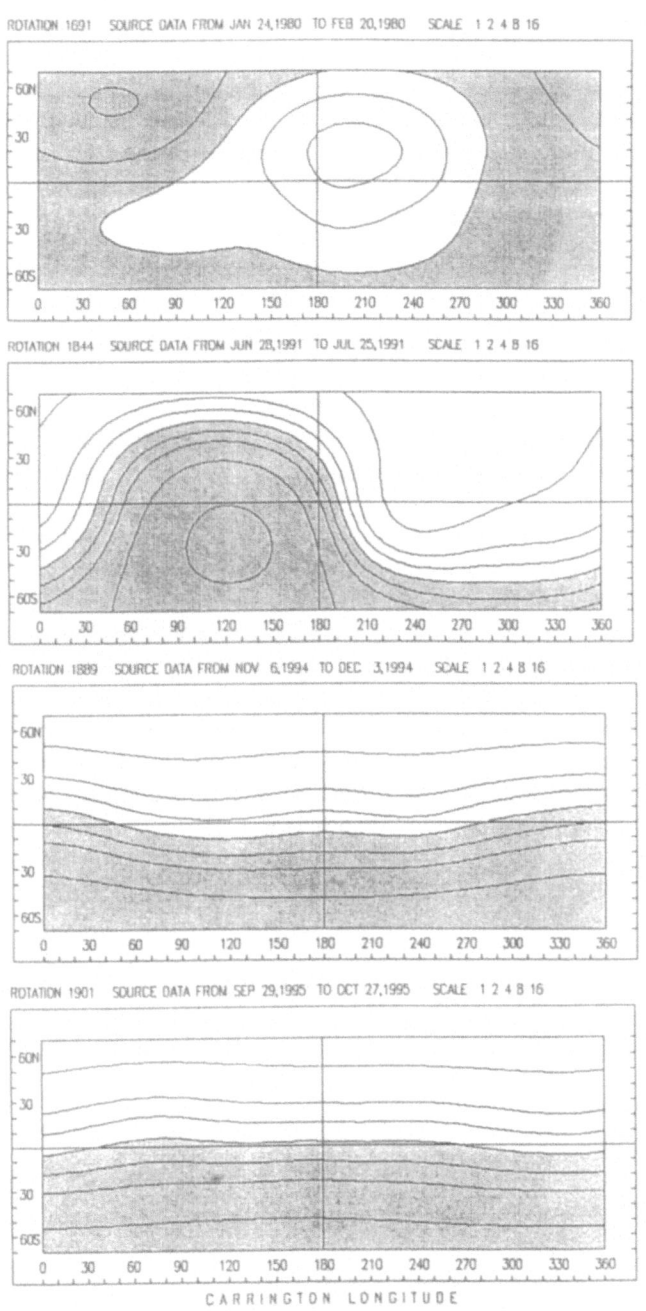

Figure 1. Global magnetic field as calculated for the source surface level 3.5R$_\odot$ for solar rotations in which eclipses on February 16, 1980, July 11, 1991, November 3, 1994 and October 24, 1995 took place. Notice that the neutral magnetic line is highly inclined to the solar equator during 1980 and 1991 eclipses (periods of the solar cycle maxima), while, it is nearly parallel to the solar equator in 1994 and 1995 (close to the solar cycle minimum). The scales are in the micro-Tesla units and the hatched areas correspond with the negative magnetic field polarity.

- In periods of the solar cycles minima (see two lower panels in Figure 1, relating to the eclipses of November 3, 1994 and October 24, 1995) the solar corona is always flattened, having clearly bipolar character [11,12]. This is in agreement with evolution of the magnetic field topology in the 22-year magnetic Hale cycle [5], because around minimum of any 11-year solar cycle the heliographic and heliomagnetic axes are more or less identical (only hemispheric polarities are reversed in the neighbouring 11-year cycles).

3. Conclusions

(a) The Ludendorff's definition of the solar corona flattening is not physically reasoned. The flattening index is strongly influenced by projection of the actual distribution of streamers onto the plane of the sky.

(b) In fact, the solar corona is probably always considerably flattened. However, this flattening is towards the solar magnetic equator and not towards the heliographic equator.

(c) Evolution of the real solar corona shapes (forms) coincides well with the long-term evolution of the solar magnetic field topology at the source surface.

(d) Cycle variations of the coronal flattening index, calculated in relation to the heliomagnetic equator, should display considerably shallower sinusoid at relatively high values of the flattening ($0.2 - 0.3$ at $2R_\odot$) than earlier presented.

(e) We suspect that one epoch of our understanding the coronal forms and their evolution with the solar cycle is over and we should inquire this phenomenon from another angle (Gulyaev [4] is also working in this direction). We are convinced that observations presently provided by LASCO C3 coronagraph at SOHO will play a decisive role in the new approach to this problem.

Acknowledgements

This work was supported by the VEGA grant 02/2007/95 of the Slovak Academy of Sciences and by the GA grant 303405 of the Czech Academy of Sciences.

References:

1. Ambroz, P. and Sykora, J.: (1994), in *Poster Papers Presented at the Seventh European meeting on Solar Physics*, G. Belvedere, M. Rodono, B. Schmieder and G.M. Simnet (eds.), Special Publication Catania Astrophysical Observatory, Catania, 121.

2. Ambroz, P. and Sykora, J.: (1994), in *Solar Coronal Structures*, V. Rusin, P. Heinzel and J.-C. Vial, (eds.) VEDA, Bratislava, 559.

3. Gulyaev, R.A.: (1992), in *Coronal Streamers, Coronal Loops and Coronal and Solar Wind Composition*, C. Mattok, (ed.) ESA, SP-348, 133.

4. Gulyaev, R.A.: (1996), *Astron. Astrophys. Trans.* (in press).

5. Hoeksema, J.T.: (1993), in *Solar - Terrestrial Prediction* - IV, J. Hruska, M.A. Shea, D.F. Smart and G. Hackman, (eds.) NOAA/ERL, Boulder, 3.

6. Koutchmy, S., Zirker, J.B., Steinolfson, R.S., and Zhugzhda, J.D.: (1991), in *Solar Interior and Atmosphere*, A.N. Cox, W.C. Livingston and M.S. Mathews (eds.), The University of Arizona Press, Tucson, 1044.

7. Ludendorff, H.: (1928), *Sitzber. Preuss. Akad. Wiss.*, **16**, 185.

8. Suess, S.T.: (1992), in *Coronal Streamers, Coronal Loops and Coronal and Solar Wind Compositio* C. Mattok, (ed.) ESA, SP-348, 63.

9. Sykora, J. and Badalyan O.G.: (1992), in *Coronal Streamers, Coronal Loops and Coronal and Solar Wind Composition*, C. Mattok, (ed) ESA, SP-348, 137.

10. Sykora, J. and Ambroz, P.: (1995), in *24th Int. Cosmic Ray Conference*, **4**, 509.

11. Sykora, J., Pinter, T., and Ambroz, P.: (1995), in *Revista da la Academia Nacional de Ciencias de Bolivia*, **69, 23**.

12. Sykora, J., Ambroz, P., Kotrc, P., Minarovjech, M., Pinter, T., Rybak, J., and Rybansky, M.: (1996), *Romanian Astr. J.*, (in press).

THERMAL EVOLUTION OF A CORONAL CONDENSATION

C. A. MENDOZA-BRICENO* and A. W. HOOD
*Department of Mathematical Sciences,
University of St. Andrews. St. Andrews,
Fife, KY16 9SS, Scotland, UK.*

Abstract. The thermal evolution of coronal magnetic structures is studied under the assumption that the inertial terms are small. Assuming the coronal heating function is due to the damping of waves, then the amount of energy supplied to the loop will decay from the footpoint towards the summit. The effect of the decay length on the heating is investigated, and when this length is below a critical value a cool condensation forms. Different initial profiles are considered and either purely hot plasmas or cool condensations can be found depending on whether the initial profile is above or below a threshold value.

1. Introduction

It is known from observations that there are many evidences of condensations in the solar corona . Prominences are an obvious example of those condensations. They consist of a cool ($T \sim 10^4$ K) and dense ($n \sim 10^{17}$) plasma embedded in the solar corona [6], and their formation still an unsolved problem. Possible mechanisms for prominences formation can be divide roughly into two categories: (a) those formed by thermal condensation [1] and (b) those formed by plasma injection [5]. Other authors have incorporated both mechanism for prominences formation [4 ,7].

In this paper we study the formation of such a cool condensation in the solar corona in terms of certain thermal equilibrium structures formed by the relaxation of a hot coronal loop-like structure due to a spatial variation of the heating function.

2. Physical Model

To begin with one assumes that the plasma is confined by a strong and constant magnetic field so that the mass flow and heat flux are essentially one-dimensional. The governing equations in dimensionless quantities are the well known equations of magnetohydrodynamics in which viscosity, resistivity and gravity are neglected, but optically thin radiation losses, thermal conduction and heating are included. Since the mechanism for coronal heating is unknown, its form will be assumed as

$$h(s) = h_* \, \exp\left(-\frac{1-s}{s_*} \right), \tag{1}$$

* Address from August 1996: Centro de Astrofisica Teorica, Facultad de Ciencias, Universidad de los Andes, Merida, Venezuela.

Z. Mouradian and M. Stavinschi (eds.),
Theoretical and Observational Problems Related to Solar Eclipses, 117-120.

where s measures the distance along the magnetic field line of symmetric loop, h_* a is energy deposition at the base of the loop and s_* is the heating scale length. The basic equations then are governed by three dimensionless parameters, namely L_* proportional to the loop length, h_* proportional to the heating deposition and s_* proportional to the spatial decay-length of the heating source.

3. Numerical Results

3.1. THE EFFECT OF VARYING THE DECAY LENGTH OF THE HEATING

The solutions for T as a function of s and time for the initial state $T(s,0)$ and boundary conditions $T(1,t) = 1$ and $\partial T / \partial s(0,t) = 0$ were explored for a broad range of values of the parameters L_*, h_*, and s_* [2].

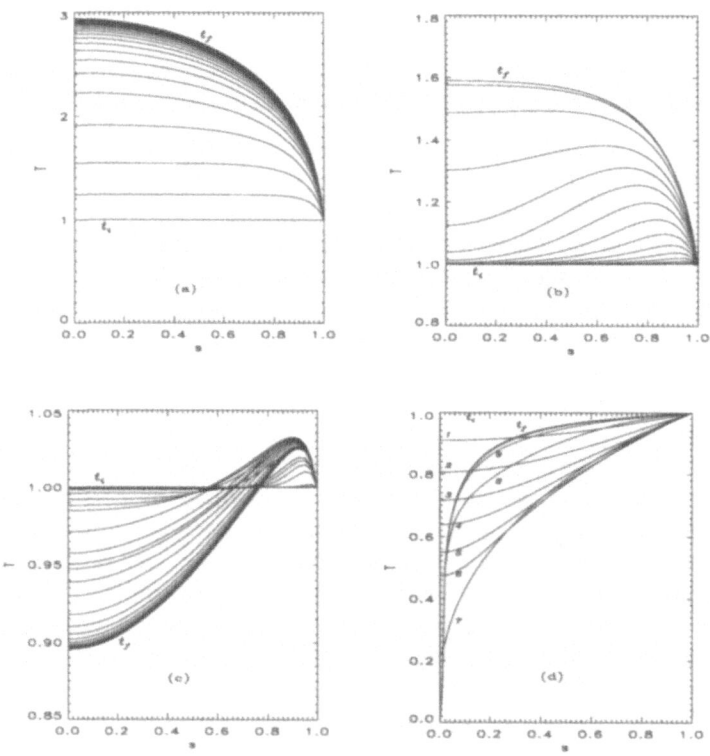

Figure 1. Temperatures profiles at selected times from t_i to t_f. For a) a large value of s_*, b) s_*=0.23, c) s_*=0.05 and d) s_*=0.01, the labelled numbers indicate intermediate times.

Figure 1 shows the temporal evolution of the temperature profile i.e. T as a function of s, for $L_* = 4.8 \times 10^{-1}$, $h_* = 1.01 \times 10^{-2}$ and for different values of s_*. These parameters correspond to a loop length of 4.5×10^7 m and a heating of 10^{-4} Wm^{-3}. The parameter s_* decreases from Figure 1a to Figure 1d. When s_* is decreased, the final temperature profile has a lower temperature at the summit.

It is also found that when $s_* < s_{*c}$ ($s_{*c} = 4.3 \times 10^{-2}$ is the critical value below which no static hot solution is found for the particular boundary conditions; cf. [3] the temperature, starting from coronal values, begins to decrease smoothly, but at a time $t = 5$ (time in dimensionless units) the temperature at the summit sharply decreases (Figure 1d) [2]. These results suggest that cool condensations can be stimulated as a result of the non-equilibrium when the energy supply is more concentrated at the footpoint than throughout the whole coronal loop.

3.2. THE EFFECT OF CONSIDERING DIFFERENT INITIAL CONDITIONS

In this section different initial conditions are taken in order to study how the temperature profile evolves to different final solutions.

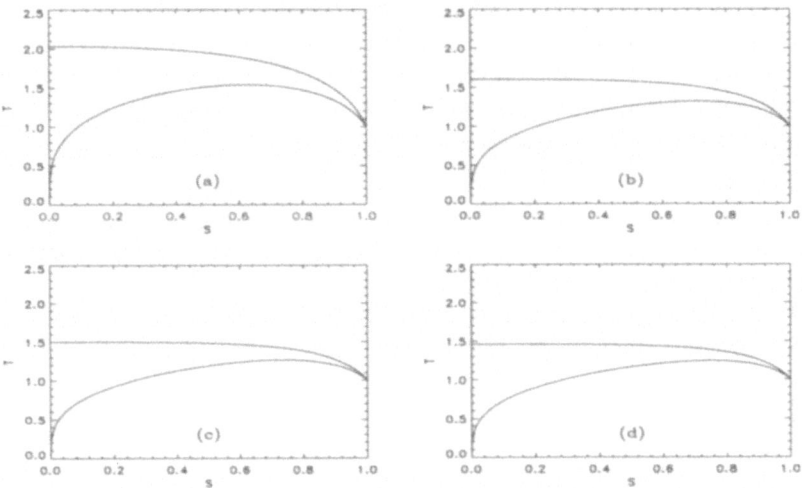

Figure 2. Final temperature profile for different values of the decay length of the heating $s_* =$ a) 0.44, b) 0.23, c) 0.20 and d) 0.19.

It is likely that two final static solutions exist for given values of the parameters L_*, h_*, and s_*. Mendoza-Briceno (1996) has shown that multiple solutions exist to the static energy equation. For example, consider an initial temperature profile of the form $T(s,0) = (T_{ch} + T_{cor} + (T_{cor} - T_{ch}) \tanh(b(s - b_0)))/2$, where T_{ch} and

T_{cor} are temperature values at the summit and the footpoints respectively, b_0 is the translation point, and b the parameter that governs the steepness of the temperature transition from T_{ch} to T_{cor}. This is not in equilibrium. However, the effect of considering the above initial temperature profile and following the evolution of the system to the final static state has been investigated by Mendoza-Briceno [2].

Figure 2 shows the final temperature profiles, after considering different initial temperature profile, for different values of s_*. These figures have been obtained using different values for T_{ch} and T_{cor} corresponding to the suitable values to allow the evolution to firstly a hot and secondly a cool summit.

It is likely there exists a threshold initial profile as an initial temperature profile in which one could say that a marginal initial temperature profile exists.

4. Conclusions

In this paper we have investigated the thermal evolution of a cool condensation in a symmetric coronal loop when the heating function depends on the spatial deposition of the heat. This form of heating function is based on the damping of waves, so the energy deposited in the loop decays from the footpoint. Its form was taken as an exponential and the evolution of the loop due to different decay lengths was studied.

It was found that the temperature at the apex decreases when the decay length of the heating is decreased up to a critical value where the decreasing of the temperature is not significantly different from coronal values. But if s_* is taken to be less than a critical value, namely s_{*c}, the temperature drops further, away from coronal values to chromospheric values.

Different initial temperature profiles were considered and it was observed that the temperature increased or decreased according to whether the initial temperature profile was above or below a threshold one.

This model stresses the importance of the form of the heating function for the evolution of a cool condensation from the hot corona, if the value of the decay length is $s_* < s_{*c}$.

References

1. Field, G.B. (1965) Thermal Instability, *Astrophysical J.* **142**, 531.
2. Mendoza-Briceno, C. A. (1996) *The Nonlinear Thermal Evolution of Coronal Structures*, PhD *Thesis*, University of St. Andrews.
3. Mendoza, C. A. and Hood, A. W. (1996) Formation of cool condensation in a magnetic structure, *Astrophysical Lett. & Communications* (in press).
4. Mok,Y., Drake, J. F., Schnack, D. D. and Van Hoven, G. (1990) Prominence formation in a coronal loop, *Astrophysical J.* **359**, 228.
5. Pikel'ner, S.B. (1971) Origin of quiescent prominences, *Solar Physics* **17**, 44.
6. Tandberg-Hanssen, E. (1974) *Solar Prominences*, D. Reidel Publ. Co., Dordrecht, Holland.
7. Wu, S.T., Bao, J.J., An, C.H., and Tandberg-Hanssen, E. (1990) The of condensation and heat conduction in the formation of prominences: An mhd simulation, *Solar Physics* **125**, 277.

MEDIUM AND LARGE SCALE STRUCTURES OF THE LOW CORONA FROM DECIMETER AND METER WAVELENGTH OBSERVATIONS

P. LANTOS
Observatoire de Paris-Meudon,
92195 Meudon cedex, France

A. COULAIS
Observatoire de Paris-Meudon,
92195 Meudon cedex, France
LPCE-CNRS,
45071 Orleans cedex 02, France

C. E. ALISSANDRAKIS
University de Ioannina,
41110 Ioannina, Greece

Abstract

Since 1991, the Nançay Radioheliograph provides daily observations of the Sun at wavelengths from 69 cm (435 MHz) to 1.83 m (164 MHz). Older observations were available mostly at 169 or 164 MHZ. Two-dimensional maps could be computed with an aperture synthesis method when the brightness distribution remains constant during the day (i.e. in absence of strong noise storms). At meter wavelengths, detected large scale structures are coronal holes, the quiet sun and a new component called "coronal plateau" which is a signature at low altitude of the coronal streamer belt and of the Heliospheric Current Sheet (HCS). Medium scale structures observed at meter and decimeter wavelengths are explained as emissions in loop systems. They may be either faint noise storm continua or thermal emissions.

1. Introduction

The observations of the quiet corona at meter and decimeter wavelengths have interesting characteristics when compare with other coronal observations. The corona, unlike in the optical range (coronal lines or K-corona) is seen on the disk as well as on the limb. On the other hand the radio emission of the quiet Sun is sensitive to lower

Z. Mouradian and M. Stavinschi (eds.),
Theoretical and Observational Problems Related to Solar Eclipses, 121-124.
©1997 *Kluwer Academic Publishers.*

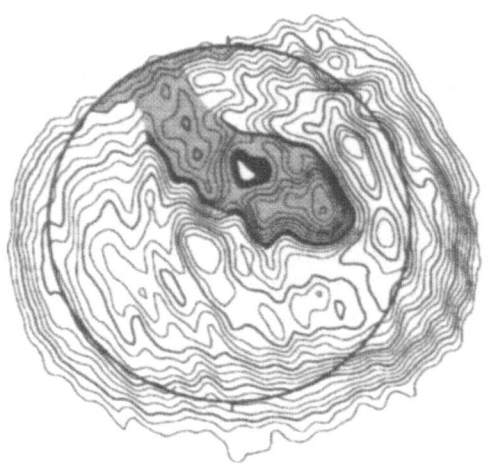

Figure 1 Map of the Sun at 327 MHz obtained on 26 May 1992 with the Nançay Radioheliograph.

densities than X or EUV on the disk. Thus radio measurements (particularly at meter or decameter wavelengths) correspond to higher altitudes than the structures seen with Yohkoh SXT or SOHO EIT instruments.

Nevertheless the radio observations, because of the absence of lines and because of the refraction are more difficult to interpret in terms of diagnostic than other coronal observations. Plane-parallel coronal models are unable to explain the observed brightness temperatures and models taking the density inhomogeneities into account are not presently available.

2. Observations

The Nançay Radioheliograph observed the Sun at 169 MHz from 1980 to 1984. After that period its frequency range was extended from 150 to 450 MHz. The instrument in operation in 1984 has been described by the Radioheliograph Group [1]. An East-West array of 18 antennas over 3.2 km and a North-South array of 12 antennas over 1.2 km observed the Sun independently at 169 MHz. The resulting spatial resolution was 1.15 arcmin in East-West and, during the summer, 4 arcmin in North-South. Since1991, the observations are at five frequencies in East-West and in North-South. The corresponding wavelengths extend from 69 cm (435 MHz) to 1.83 m (164 MHz), the spatial resolution being roughly proportional to the wavelength.

Two-dimensional maps have been obtained with the aperture synthesis method based on Earth rotation [2,3]. The main limitation of this method is the assumption of stable brightness distribution during the 8 hours of daily observation.

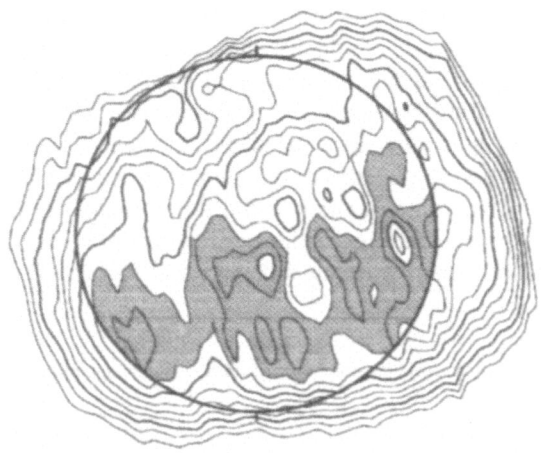

Figure 2 Map of the Sun at 164 MHz obtained on 26 May 1992 with the Nançay Radioheliograph.

When intense noise storms are present it is not possible to compute a synthesis map, because they vary in brightness.

3. Comparison of Maps at Decimeter and at Meter Wavelengths

Figure 1 shows a map obtained at 327 MHz (decimeter range) with the Nançay Radioheliograph on 26 May 1992. Figure 2 shows the map at 164 MHz (meter range) on the same day. Outer isophote corresponds to 250,000 K and other isophotes are drawn with steps of 50,000 K. On figure 1 a coronal hole is seen as depression in brightness temperature, as shows by the grey area. The lower brightness temperature measured in the coronal hole is 400,000 K. In the middle of the coronal hole an isolated active region, left in white, is seen as a faint emission source 100,000 K brighter than the surroundings. This region, as well as the coronal hole are seen at the same location as in X-ray image taken by Yohkoh SXT experiment. On figure 2 (164 MHz) the depression regions are more scarce. The lower brightness temperature of the coronal hole is at about the same location as at 327 MHz and corresponds to 600,000 K. On figure 2 the region indicated in grey is a large scale emission, called coronal plateau [4,5], surrounding the emission sources (the brightest are left in white). Its brightness temperature is about 100,000 K above the quiet sun level. The emissions sources seen at 164 MHz are generally not located above active regions where the loop density is too high. They are frequently related with larger scale loop systems outside faculae [6].

The brightest emissions at 327 MHz, located close to the West limb are non-thermal emissions (noise storm continua) and have brightness temperatures above 1,300,000 K. The northern one has no counterpart at meter wavelength and is close to a sunspot group. Other emissions sources at that wavelength are, unlike emissions sources at 164 MHz, generally located above faculae. This explains important differences between maps at decimeter and at meter wavelengths. Radio maps at decimeter wavelengths are rather similar to the X-ray images obtained with Yohkoh SXT instrument [7] while the observations at meter wavelengths give additional information on large scale loops with lower density and higher altitude.

4. References

1. The Radioheliograph Group, (1983), The Mark III Radioheliograph, preliminary results, *Solar Physics*, **88**, 383.

2. Alissandrakis C.E., P. Lantos, E. Nicolaïdis, (1985), Coronal Structures Observed at Metric Wavelengths with the Nançay Radioheliograph, *Solar Physics*, **97**, 267.

3. Coulais A., (1996), Thèse, Université Paris VII.

4. Lantos P., C.E. Alissandrakis, (1991), Detection of the coronal neutral sheet on the solar disk, in E. Marsch and R. Schwenn (ed.), *Solar Wind Seven*, Pergamon Press, Oxford, p 213-216.

5. Lantos P., C.E. Alissandrakis, D. Rigaud, (1992), Quiet Sun and emission sources at meter and decimeter wavelengths and their relationship with the coronal neutral sheet, *Solar Physics*, **137**, 225.

6. Lantos P., C.E. Alissandrakis (1996) in preparation.

7. Lantos P., C.E. Alissandrakis and the YOHKOH Team, (1995), Large Scale Structure of the Solar Corona Observed at Meter and Decimeter Wavelengths and in X-rays, *Advance in Space Research*, **17**, 261.

FILAMENT CHANNELS IN THE CORONA

O. ENGVOLD
Institute of Theoretical Astrophsics
University of Oslo, Oslo, Norway

Abstract Filament channels, where filaments (prominences) are formed, are relative long-lived, narrow lanes between extended areas of magnetic field of opposite polarity. The channels are characterized by arcades of magnetic loops overlying an inner low-density cavity. The talk will review the observational bases, including small- to large-scale structure and dynamics, showing how filament channels differ from "normal" corona. One may subsequently understand better the conditions that lead to formation of filaments.

1. Introduction

Understanding the physics of filament channels is relevant for understanding the solar corona, as well as solar prominences (filaments). Filament channels are fundamental structures and prerequisite for the formation of filaments. They exist both before and after the appearance of filaments, i.e. they may exist without containing any observable filament structures. Channels are identified in the photosphere as the corridors of magnetic polarity reversal, in the chromosphere by long zones of aligned Hα fibrils, and in the corona as low density cavities underneath arcades of bright loops straddling the opposite magnetic polarities in the photosphere below.

What does observations tell about how filament channels are formed, their magnetic fields and plasma, and what can be inferred about conditions for formation of low temperature filament plasma? The talk will discuss briefly some of these questions.

2. Observations

2.1. FILAMENT CHANNELS OBSERVED AT ECLIPSES

Eclipse observations show that quiescent prominences are embedded in a complex coronal structure of a darker cavity within arches at the base of thin coronal streamers (Saito and Tandberg-Hanssen 1973). The outer helmet streamer structure is often seen in white-light eclipse observations.

A unique set of high resolution, white-light images of the solar corona were obtained with the 3.6 m Canada-France-Hawaii Telescope on Mauna Kea during the eclipse of July 11, 1991 [15, 25]. The processed images show arcades of concentric, fine-scale, dark and bright loop-like structures extending above a small prominence at the limb. A small coronal cavity close to the prominence contains numerous fine, dark

Z. Mouradian and M. Stavinschi (eds.),
Theoretical and Observational Problems Related to Solar Eclipses, 125-130.
©1997 *Kluwer Academic Publishers.*

threads which demonstrate better than any data before that a cavity region is highly structured and inhomogeneous.

2.2. X-RAY AND EUV OBSERVATIONS ON THE DISK

With the advent of normal incidence imaging (NIXT) of the corona [8], a new level of spatial resolution could be obtained. Gaizauskas et al [7] analyzed coronal (NIXT) observations of a quiescent prominence at the SW limb of the Sun during the total solar eclipse of July 11, 1991, and noticed weakly enhanced soft X-ray emission that coincided with an Hα bright rim [11]. The enhanced X-ray emission is evidence for magnetic reconnection on a small scale and it follows that this may also be the cause of Hα bright rims of quiescent prominences. The technique of normal incidence multilayer-coated optics is presently used with great success in the new "Extreme-Ultraviolet Imaging Telescope" (EIT) for the SOHO Mission [2]. The powerful combination of several additional instruments on board SOHO, such as, "The Coronal Diagnostic Spectrometer" (CDS) [10], and SUMER ("Solar Ultraviolet Measurements of Emitted Radiation")[33] is expected to play a central role in studies of the solar corona, including filament channels, in years to come. Very recent observations with SUMER and CDS show noticeably reduced EUV line fluxes at the location of filaments [6]. The lower line fluxes agree with earlier EUV Skylab observations of prominences at the limb [13]. Hence, the Prominence-Corona Transition Region (PCTR) contains less emitting mass than the corresponding Chromosphere-Corona TR.

Data from the Japanese Yohkoh satellite observatory [32] has provided new and fundamental insight on the structure and dynamics of the solar corona. High resolution soft X-ray images from Yohkoh shows bright coronal loops overlying magnetic neutral lines containing dark Hα or He I 10 830Å filaments. Such loops are evidently part of the helmet arch structures seen in eclipse photographs [29].

Pojoga [26] studied the variation with height of the Emission Measure (EM) from EUV lines and concluded that the emitting structures appear to have sub-resolution geometries. An inferred small-scale structure of the PCTR would also imply a highly inhomogeneous cavity. A similar conclusion was reached by Dere et al [3] for the CCTR, from comparison of column mass densities and densities derived from density sensitive line ratios. They found that the volume fill factor of the CCTR plasma must be less than 0.01.

2.3. EVIDENCE FOR HIGH AND LOW TEMPERATURE GAS IN CORONAL CAVITIES

From the thermal broadening of coronal Fe XIV emission lines Bessey and Liebenberg [1] concluded that the temperature is the same inside and outside cavity regions. It is generally accepted that the lower brightness of coronal cavities is due to a lower plasma density [30].

Leroy[17] found evidence of very faint Hα emission in the corona close to prominences. The same data was re-examined by Dermendjiev et al.[4] who measured faint coronal Hα emission also in the form of long systems of drops and fibrils outlining large, smooth curves that one normally associates with magnetic loops. The significance and consequences of the presence of faint, low temperature emission close to prominences, is that coronal cavities also contain a noticeable amount of low temperature gas.

The fact that Hα filaments occupy only small fractions of a filament channel [30] is not a variance with the idea that a larger volume of the cavity may contain low temperature gas. The visibility of cold matter, as observed in a strong chromospheric line like Hα, depends on both line opacity and source function of the absorbing matter [5]. Also, studies of radio filaments [9] suggest that there are much more "cold" material in filament channels than one detects in Hα.

2.4. STRUCTURE AND DYNAMICS

Martin and Echols [21] derived the direction of the axial fields of channels from the orientation of neighboring Hα plagettes, associated fibrils, and knowledge of the polarity of the magnetic field in the photosphere on either side of the filament. When the field is directed to the right, when viewed from the positive polarity side of the inversion line, the filament channel is labeled "dextral". In the opposite cases it is said to be "sinistral". Filaments themselves also fall into two categories, "dextral" and "sinistral", recognizable by the systematic orientation of the *barbs* along their two sides. Viewed from above and along its axis in the direction of the magnetic field of the channel, a filament is dextral when the barbs make right-hand turns away from the filament axis. A filament is sinistral when the barb make left-hand turns away from the filament axis. Dextral filament channels are always associated with dextral filaments, and vice versa. A most remarkable result from the studies of Martin *et al.* [20, 21], is that the orientation of channel fields, in the cases of quiescent filaments, differs systematically in the northern and southern hemispheres. There are exceptions to the rule, but about 80% of all quiescent filaments obey the dextral/sinistral precedence.

Martin and McAllister [22] studied the orientation of X-ray arcades over filament channels from Yohkoh data, and they noticed a change in magnetic shear in altitude. The shear may be quantified as a "skew angle", or acute angle, between a tangent to the top of a bright loop and the long axis of the filament or its channel. The observed change in skew angle from low to high in the corona can be either "right-handed skew" or "left-handed skew". The authors find that dextral filaments are associated with "left-handed skew" and sinistral filaments with the opposite skew.

It follows that filament channels represent a highly organized large-scale system of magnetic arcades that are anchored in the photospheric network flux. It is not yet established, observationally, whether or not photospheric flows actually contribute to the formation and maintenance of the channels. Martres *et al* [23] showed that the line-of-sight velocity amplitudes in the photosphere underneath filaments were smaller than average. Also, Hα filaments and the polarity inversion lines coincide with the zero velocity line between regions of respectively up- and downward directed motions. These results are later confirmed by Iospha and co-workers [12].

Filament channels lying above the magnetic polarity reversals are subjected to stretching by solar differential rotation. The channels are otherwise remarkably stable, in spite of the continuous dispersive action of the supergranulation flow field of the photosphere [14, 28] measured large-scale meridional circulation flows which embodied bands of up-flows and down-flows that appeared to be co-spatial with long-lived filament channels. The occurrence of filaments seemed to be a factor $1 \, ^1/_2$- 2 higher near up-drafts as compared to down-drafts. It remains to be shown whether or not these slow (a few m/s) flows participate in the formation of filament channels.

Horizontal motion of converging magnetic fields has frequently been observed in connection with filament formation [19], but there is no clear observational evidence for systematic large-scale photospheric flows that may account for the formation of channels. Using the technique of "Local Correlation Tracking", Molowny [24] failed to detect any systematic, large scale converging or, oppositely directed (sheared) flows, on the two sides of the polarity inversion line. However, in order to detect a possible low velocity flow, one will evidently have to suppress the much stronger velocity signal from the supergranulation flows by a suitable temporal and spatial averaging of the data.

The noted variation in skew angle of the observable structures, from a filament to the associated coronal arches above [22], is puzzling. It has been noticed by Martin and co-workers that filament barbs seem to connect to the minority polarity flux on either side of the polarity inversion line [18], whereas coronal arches appear to be rooted in the network flux (majority polarity). This implies that here may be a layer of sheared magnetic fields, at some height in the cavity, where the magnetic spine of the filament touches the lower boundary of the helmet structure. One notices, occasionally, in Yohkoh images the presence of faint, thin filamentary structures that run close to Hα filaments [31]. Such faint X-ray filaments might possibly be associated with sheared magnetic layers, where local heating may take place, in coronal cavities.

3. Modeling of Filament Channels

A magneto-static flux rope model, that accounts for the inferred relationship between prominences and its coronal cavity, has been put forward by Low and Hundhausen[16]. Priest *et al.* [27] have developed a dynamic model that accounts for the organization of dextral and sinistral filaments and filament channels by combining the effects of differential rotation on subsurface flux, of magnetic buoyancy to flux as a filament channel with magnetic field oriented along its axis.

The observational studies by Martin *et al.*[20, 21] suggest that the magnetic field of quiescent filaments is in the form of long and flat (non-potential) flux tubes. The taller, loop-like magnetic arcades of the overlying helmet structures that extends downwards into the cavity, may serve to suppress and flatten the magnetic field of the associated filament. Theoretical investigations of such configurations are encouraged.

4. Concluding Remarks

The physical conditions in a coronal cavity, its magnetic structure and low density plasma, are still poorly understood.

It seems unlikely that a depletion of coronal gas in cavities is the result of condensation of coronal matter into much denser, low temperature filament gas. A more probable reason for the lower density could be that a coronal cavity in pressure balance with its surroundings will hold a slightly higher magnetic pressure than the normal corona. Also, cavities differ from typical corona by containing a component of low density, low temperature plasma. Coronal cavities are presumably highly inhomogeneous, as also suggested from eclipse observations by November and Koutchmy [15]. Further theoretical investigations are needed to fully comprehend the complex magnetic filament channels and their low density, high and low temperature cavities.

The physics of solar prominences and their formation will remain elusive until the physics of filament channels, and their magnetic structure and puzzling cavities, are better understood. A closer study of the fine-scale structure and low temperature plasma of filament channels and coronal cavities could be a suitable and important observational task for the total solar eclipse of 11 August 1999.

References

[1] Bessey R J, and Liebenberg D H: (1984) *Solar Phys* **94**, 239
[2] Delaboudiniére J-P, *et al.*: (1995) *Solar Phys* **162**, 291
[3] Dere K P, Bartoe J-D F, Brueckner G E, Cook J W, and Socker D G: (1987) *Solar Phys* **114**, 223
[4] Dermendjiev V N, Mouradian Z, Duchlev P, and Leroy J-L: (1994) *Solar Phys* **149**, 267
[5] Engvold O: (1994) in *"Solar Coronal Structures"*, IAU Coll.144 V Rusin, P Heinzel and J-C Vial, (eds.) VEDA Publ. Company. p. 297
[6] Engvold O, et al.: (1996) (In preparation)
[7] Gaizauskas V, de Luca E, Golub L, and Jones H P: (1994) *BAAS* **26**, 1522
[8] Golub L, et al.: (1990) *Nature* **344**, 842
[9] Gopalswamy N, White S M, and Kundu M R: (1991) *Astrophys. J.* **379**, 366
[10] Harrison R A *et al.*: (1995) *Solar Phys* **162**, 233
[11] Heinzel P, Kotrc P, Mouradian Z, and Buyukliev G T: (1995) *Solar Phys.* **160**, 19
[12] Iospha B and Kulikova E: (1993) in *"International Conference on Solar Magnetic Fields"*, M. Schüssler and W. Schmidt(eds).. 29 June - 2 July, 1993, Freiburg,. 387
[13] Kjeldseth-Moe O, Cook J W, and Mango S A: (1979) *Solar Phys.* **61**, 319
[14] Komm R W, Howard R F, and Harvey J W: (1995) *Solar Phys* **158**, 213
[15] Koutchmy s et al.: (1994) *Astron. Astrophys* **281**, 249
[16] Low B C and Hundhausen J R: (1995) *Astrophys. J* **443**, 818
[17] Leroy J-L: (1972) *Solar Phys* **25**, 413
[18] Martin S F: (1994) *BAAS* **26**,1522
[19] Martin S F: (1986) in *"Coronal and Prominence Plasmas"* NASA Conf. Publ. 2442, A I Poland. (ed.) 73
[20] Martin S F, Bilimora R, and Tracadas P W: (1993) BAAS, **25**, 1217
[21] Martin S F and Echols C R: (1994), in *"Solar SurfaceMagnetism"*, RJ Rutten and CJ Schrijver. (ed.) Kluwer Academic Publ. 339
[22] Martin S F and McAllister A H: (1995) *BAAS* **27**, 961
[23] Martres M J, Rayrole J, and Soru-Escaut I: (1976) *Solar Phys* **46**, 137
[24] Molowny Horas R L: (1994) Dr.Sci. Thesis Univ. Oslo
[25] November L J and Koutchmy S: (1996) *Astrophys. J* **466**, 512

[26] Pojoga S: (1994) in *"Solar Coronal Structures"*, IAU Coll.144 V Rusin, P Heinzel and J-C Vial, (eds.) VEDA Publ. Company. 357

[27] Priest E R, van Ballegooijen A A, and MacKay D H: (1996) *Astrophys. J* **460**, 530

[28] Ribes E, Mein P and Mangeney A: (1985) Nature **318**, 171

[29] Saito K and Tandberg-Hanssen E: (1973) *Solar Phys* **31**, 105

[30] Serio S, Vaiana G S, Godoli G, Motta S, Pirronello V, and Zappala RA: (1978) *Solar Phys* **59**, 65

[31] Solberg F C and Engvold O: (1996) (In preparation)

[32] Tsuneta S, Takahashi T, Acton L W, Bruner M E, Harvey K, andOgawara Y: (1992) *Pub. Astron. Soc. Japan* **44**, L211

[33] Wilhelm K *et al.*: (1995) *Solar Phys* **162**, 189

FAINT "CHROMOSPHERIC" EMISSION IN THE SOLAR CORONA: OBSERVATIONAL AND THEORETICAL PROBLEMS

V. N. DERMENDJIEV

Institute of Astronomy, Bulgarian Academy of Sciences, 72
Trakia Blvd., 1784 Sofia, Bulgaria

1. Introduction

A total eclipse offers a rare opportunity for ground based observers within the shadow path to study the less luminous solar corona at the moment when the intensely bright photosphere is occulted by the Moon's disk. The free electrons that scatter photospheric radiation producing the K corona are concentrated usually in the regions with strong magnetic fields forming such coronal features as streamers, helmets, plumes, and polar rays. The coronal spectrograms reveal a great number of coronal emission lines (see for example [1]) over the whole range of the spectrum from $\lambda6900$ to $\lambda3100$ Å. Their existence is attributed to forbidden transitions of very highly ionized atoms of iron, nickel, and calcium, which have lost from 9 to 14 of their electrons under extremely low pressure and high kinetic temperature. Besides the usual features of the corona some observers report about unexpected, usually faint, low excitation emissions at the wavelength of the K and H Ca II, D3 He, Hα and others, registered higher in the solar corona.

Are such low excitation of emission lines, observed in the eclipse coronal spectra, of coronal origin or they are simply: 1) primary or secondary scattered light from prominences or from unobscured chromosphere soon after the second contact; 2) emission due to diffuse prominence matter high in the corona; or 3) instrumental effect?

If these emission lines are of coronal origin, then the questions arise: are they due to a real local transitory phenomenon observed by chance, or they are general feature of the corona? From such a point of view the problem for the existence of such chromospheric like emissions from the solar corona is intimately connected with the fundamental problem of mass and energy balance in the solar atmosphere.

In my review talk I will make a survey on the observational results having relation to this low excitation emission. I will comment the basic objections on the coronal nature of these emission lines and will discuss on the main arguments against the chromospheric and prominence origin of the faint Ca II and H I emissions from the corona. I will discuss as well as two other questions: first, what the theory could say about the possibility of generation of such emission, and second, what the theory could

Z. Mouradian and M. Stavinschi (eds.),
Theoretical and Observational Problems Related to Solar Eclipses, 131-139.
©1997 *Kluwer Academic Publishers.*

advise us in searching observational evidences in low excitation emissions from the corona.

2. What Can we Learn from the Observations?

The existence of a chromospheric like emission from the solar corona has been widely debated among the solar physicists as early as the second half of the last century. Many visual observations made at eclipses in this time support the supposition of existence of low excitation emission lines in the solar coronal spectrum. I will mention two of them - those of Rayet and Janssen (see [2]). Rayet at the eclipse of 1868 noticed that the D line, $\lambda5303$ Å and H_β are extended to greater heights than the other prominence lines. On the side of the spectrum corresponding to the Moon's limb all the lines were sharply cut off. At the 1971 eclipse Janssen traced the hydrogen spectrum to 10 arcmin above the solar limb with a radial slit. This extension of the hydrogen lines into the corona beyond the prominences he ascribed as indicative of genuine hydrogen emission from the corona. The condition of high transparency in which those observations were made, facilitated also the discovery of the absorption lines in the outer corona that rise from the F-corona.

At the beginning of this century the concept was changed. The registered H and K Ca II line in the spectroheliograms were attributed to large prominences [3] or regarded as produced by chromospheric light scattered into Earth's atmosphere from the edge of the totality umbra (for a review see [4]). Later on, in the second half of this century the discussion on the nature of this emission flared again. Colacevich [5] during the 1952 eclipse registered H and K Ca II and H_γ at a distance 70 arcmin from the Moon's limb. He put the hypothesis that this emission is of coronal origin and is due to diffuse prominence matter that could be a local phenomenon observed by chance, or general feature of the corona. Other authors adduced arguments for noncoronal origin of such low excitation emission. Migeotte and Rosen [6] explain the observed H and K Ca II emission in the outer corona as due to a diffusion in the Earth's atmosphere of the light of prominences incompletely occulted during the first seconds of the totality. Parijskij et al. [7] and Caccin et al. [8] attributed it to primary scattered prominence light in the Earth's atmosphere (aureole effect). Some years later Stellmacher and Koutchmy [9] interpreted the observed emission in H_δ, H_γ, H_β and H and K Ca II as double scattered light of chromospheric origin in the Earth's atmosphere. They showed that most of the single (or primary) scattered light is due to the illumination of the terrestrial atmosphere, by light coming from the chromosphere, concentrated in a region 40 km wide outside the eclipse path umbra. During the 1970 solar eclipse Gurtovenko and Alikayeva [10] registered at heights 4×10^4-10^5 km above the solar limb low excitation emission not of instrumental origin and not directly connected with an active prominence of the tornado type. The large scale picture of the corona obtained at the same time reveals in this area a system of thin streamers, 10 times denser than the surroundings. Later on, Alikayeva [11] reported more concretely that hydrogen, helium and H and K Ca II emissions appear in the vicinities of prominences.

New arguments in support of the coronal origin of the observed low excitation emission were given by Bappu et al. [2]. During the 1972 eclipse they registered by a slit spectrograph, covering a range of solar radii from 1.0 to 1.5 $R_\odot$ and nearly 90° in position angle, some lines (H_γ, H_ξ, H_η) from the Balmer series, D3, H and K Ca II. The emission had a maximum near 1.5 $R_\odot$. The authors supposed that a cool column shape component in the outer corona could be coronal source of these emission lines.

The airborne observations of the coronal spectrum at solar eclipses afforded new opportunities to study the problem for the low excitation emission in the solar corona. Such observation have the advantages for good seeing conditions and that an observer could remain in the total eclipse shadow longer time. The latter circumstance permit more exposures to be taken. Interesting results were obtained during the 1963, 20 July solar eclipse [4, 12] on the board of the APEQS DC-2 jet airplane flying at a height of 11 000 m by spectrograph with a slit radially set along the equator and HNB polarizer behind the slit. H and K Ca II emission lines of coronal origin were registered at a height $1R_\odot$ above the limb. They were attributed to the resonance scattering from Ca II ions that occur in a relatively cool cloud immersed in the corona. In [12] this emission is considered as transitory feature of the solar corona originating in a large structure whose size is comparable to a normal coronal hole with temperature some order of magnitude lower than the typical coronal temperature to allow the presence of Ca II ions. The slow moving electrons in it must scatter a Fraunhofer spectrum which is added to the spectrum of the F-corona. In this case the observed spectrum is the sum of three spectra: 1) continuum spectrum due to the K-corona; 2) Fraunhofer spectrum due to the F-corona and 3) also Fraunhofer spectrum scattered by the slow moving electrons of the cold region.

Conway et al. [13] took also the advantages of the airborne observations and with two spectrographs on a CI35 aircraft covering the wave length region 3 000 - 6750 Å succeeded to take six exposures. The obtained limb-to-corona intensity profiles of the observed lines revealed an unusual profile of the Ca II line in which Ca II reappears at $1R_\odot$ above the limb.

The obtained results during airborne observations, especially these of Conway et al. [13] had stimulated many solar astronomers to search with the aid of interference filters for emission in the corona in H_α, H and K Ca II and D3 lines. During the 1970 eclipse Evans [14] observed with interference filters 25 Å wide at half maximum for studying the morphology of emission objects in H_α and H and K Ca II. He concluded that no structural features larger than 25 arcsec appear in the exposure of 150 sec beyond ~ 0.6 $R_\odot$ above the limb. Recently Kanonovich et al. [15] reported for Ca II emission in the lower corona registered by 3933 Å birefringent Halle filter with 0.5 Å passband and exposure time 100 sec during the 1981 eclipse. The measured intensity is by 1-2 orders of magnitude larger than the continuum emission in the same wavelength interval, and with larger gradient.

More than forty years ago Öhman [16] proposed to use over-exposed photographs made by coronagraph for registration of very faint H_α emission in the vicinities of quiescent prominences. Leroy [17] applied this method and studied its possibility and frontiers of use. He registered by coronagraph and H_α Lyot filter of Pic-du-Midi Observatory definite structures in faint H_α emission often detected down to the threshold 2×10^{-6} times of the Sun's brightness. Leroy did not find clear relation of these structures to the low-level ordinary prominences. In his comments on Leroy's paper Öhman [18] draws attention to the probable connection of this emission with faintly luminous H_α obscuring prominences which appear sometimes as dark lanes and markings in ordinary prominences. Concrete study of the relation of the faint H_α emission in the corona to the quiescent prominences was performed in [19] on the base of Leroy's observations. It is quite possible such emission structures to relate with coronal plasmoid events such as this one observed during the 1991 total solar eclipse by Canada-France-Hawaii telescope with high speed photographic and video CCD cameras (see [20]).

Recently Dermendjiev et al. [21] made the conclusion on the base of Leroy's observations that these emission structures resemble systems of drops and fibrils outlining curves similar to those which are generally supposed for the prominence magnetic field configuration. As well, almost every object of faint $H\alpha$ emission is located at a place where there is a local increase of the green coronal emission irrespective of its brightness.

At the end of this section I will remind a very curious episode in the study of the problem of existence of low excitation line emission in the corona. Gnevishev and Gnevisheva [22] reported that they have observed from June 1954 to March 1955 at Kislovodsk Coronal Station intensive helium emission in the line D3 ($\lambda5875.6$ Å) high above the chromosphere, mostly between the prominence arches inclined one to the other, even when the distances between the prominences were large. These observations had stimulated many discussions and observational studies on the problem. Very soon, however, Zirin et al. [23] found that this emission appears on spectrograms as a results of chromospheric emission scattered by secondary optics of the coronagraph. On the under occulted spectrograms, in which chromospheric light could reach the secondary optics, D3 line was invariably observed.

3. What Can we Learn from the Theory?

It is surprising that the corona at a temperature of over 10^6 K should emit low excitation (or chromospheric) emission lines. The origin of this emission remains still rather uncertain. A view of the articles on this topics illustrates the controversial state of the interpretations concerning this problem. The common conclusion of the authors studying such emission during solar eclipses is that the appearance of "cool" emission in the corona seems to be connected with short-living transitional active processes that occur in the corona. Some of the authors [24, 4, 12] conclude more concretely that this emission may be due to temperature variations in the corona. I will mention again the proposed physical ideas for interpretation of the coronal origin of the observed low excitation emission: 1) diffuse prominence matter in the corona, that could be local phenomenon observed by chance or general feature of the corona [5]; 2) cool "clouds" immersed in the corona [4]; and 3) a cool column shape component in the outer corona [2].

However, the basic question, namely, what kind of mechanism will produce such lower temperature structures in the hotter solar corona, remains without answer. It seems to me that the contemporary MHD theory and the theory of plasma instabilities suggest possibilities for a new treatment of the problem. They could create a ground for theoretical arguments if we need motivation for future observational studies of this problem. On the other hand, however, it is very difficult observational problem to determine the temperature structure on a fine scale, since a high geometrical resolution is required for the observation of weak coronal lines.

Concerning the basic question, namely, what kind of mechanism could decrease the electron kinetic temperature in a short time and in small scale, I will try to outline the possible ways for a concrete theoretical treatment of the problem, mainly based on the possible plasma effects due to the coronal magnetic field changes and due to plasma motions.

Let we trace back the basic supposition concerning the physical conditions in which faint emission of hydrogen, helium and metals is possible. According to Shklovskij [25] the equivalent widths of some lines of the Balmer series of the solar spectra at a distance $1.2 \, R_\odot$ (Table 1) are so small that these emission lines could be observed if the electron temperature in the solar corona become less than 10^5K. Alikayeva [11] has obtained the following values for the basic physical parameters of an observed emission structure near to a tornado prominence (Table 2).

TABLE 1. Equivalent widths at a distance $1.2 \, R_\odot$ according to [25]

	Electron temperature [K]					
	10^4	2×10^4	4×10^4	8×10^4	1.6×10^5	3.2×10^5
H_α	1.1	0.57	0.37	0.17	0.07	0.03
H_β	0.47	0.26	0.16	0.06	0.03	0.01
H_γ	0.25	0.13	0.08	0.03	0.01	0.005

TABLE 2. Physical parameters of the faint emission region according to [11]

Height	h = 75″		h = 80″		h = 85″	
n_e [cm^{-3}]	10^9	10^{10}	10^9	10^9	10^9	10^{10}
T_e [K]	2.2×10^4	1.2×10^4	3×10^4	1.7×10^4	2.5×10^4	1.1×10^4
L [cm]	1.4×10^9	5.0×10^6	4.3×10^9	2.0×10^7	1.3×10^9	4.2×10^7

Leroy [17] considers that there are three cases in which generation of faint H_α emission is possible (Table 3), depending crucially on the density n, geometrical dimension L and the temperature T.

TABLE 3. Physical parameters of faint H_α emission structure according to [17]

n [cm^{-3}]	L [arcsec]	T [K]	Remarks
10^{10} - 10^{11}	1	5×10^4	Emission in UV
10^{10} - 10^{11}	0.26	10^4	Very small
$< 10^{10}$	1	10^4	not in the very low corona

The analysis of these Tables leads to the conclusion that the supposed physical conditions in which generation of low excitation emission is possible rarely occur in the solar corona. This is the basic difficulty.

Öhman [26] paid attention on the possible role of the magnetic filed in the process of formation of regions of low excitation in the prominences and noted the possibilities which give the magnetic cooling mechanism, proposed by Marshall [27] for explanation of coronal streaming phenomena in sunspot type prominences. The basic idea of this mechanism was developed in [21] in a manner to be suitable for interpretation of the process of generation of faint $H\alpha$ emission in the corona. The required change in the kinetic temperature T_{kin} could be produced by the electric field **E** induced during a change in a magnetic field **H**. The relative energy change $\Delta W/W$ of the electrons will be proportional to the relative magnetic field change $\Delta H/H$ according to the relation

$$\frac{1}{t}\frac{\Delta W}{W}\frac{1}{H}\frac{\partial H}{tt} \approx \frac{1}{\Delta t}\frac{\Delta H}{H}, \tag{1}$$

where $t = 2\pi mc/eH$ is the period of a complete cycle of an electron around a magnetic line of force, Δt is the time interval of the magnetic field perturbation, and $W = \rho^2/2m$ is the electron energy: m, e, ρ are the mass, charge, and momentum of the electron, c is the light velocity.

In other words, the changing magnetic field has the effect of applying an electric field on electrons and protons, which are spiralling along the lines of force, sometimes in such a way that their velocities decrease that means that the plasma T_{kin} decreases. In such a case because the electrons become slower, their collisions with the protons increases resulting in the formation of hydrogen atoms (recombination) and a "cloud" of "condensed" atoms appears. Its morphology depends on the local ratio of $\Delta H/H$ and the configuration of the local magnetic field. The main effect is the formation of a cloud of hydrogen atoms in a region where previously there were mainly protons and electrons. At the same time, the total mass density does not change and no matter is required to cross the lines of force. The quantitative estimates obtained in [21] show that for an altitude $h = 10^5$ km in the corona the corresponding cooling becomes $\Delta W = 0.68$ erg cm^{-3}, for a cooling time $\Delta t = -1.3 \times 10^{-10} \, 1/H(\Delta H/H)$, when the magnetic field undergo a perturbation of decrease. For the most probable coronal conditions $H = 1\div0.1$ G and $\Delta H/H = 1 \div 10$ % the cooling time is $\Delta t = 10^{-10} \div 10^{-12}$ which means that the proposed cooling process could be considered as an adiabatic expansion. It is shown in [21], as well, that the volume of the cool matter is $\approx 10^2$ times greater than that of the hot one and the gas pressure of this matter is $\approx 1.5\times10^{-6}$ bar.

Another point of view was proposed in [28] where the faint $H\alpha$ emission was regarded as result of development of plasma instability in presence of slow waves carrying negative energy. According to Sturrock [29] for a wide class of dynamical systems slow waves carry negative energy. In an one-dimensional model the slow waves propagating in a moving medium refer to such waves whose phase velocity V_{ph} changes its sign on transforming from a moving to a stationary frame. As well, as it was shown by Kodomtsev et al. [30], a nonequilibrium transparent media may posses anomalous dispersion and in this case the energy of a monochromatic electromagnetic wave may be negative in the sense that the energy of a medium with a wave is less than the energy of a medium in the absence of the wave. For the case of longitudinal oscillation ($H = 0$) and isotropic transparent medium the energy of a monochromatic electromagnetic wave is

$$U = \frac{1}{8\pi}\left[\frac{d}{d\omega}(\varepsilon\omega)\langle E\rangle^2\right].$$ (2)

It will be negative if the dispersion $\dfrac{d\varepsilon}{d\omega} < 0$, ε is the dielectric permeability and ω is the electromagnetic wave frequency. For a collisionless coronal plasma with "cool" electrons and anisotropic velocity distribution function of the ions, near to the n-th harmonic of cyclotronic frequency of the ions $\Omega_i = eH/m_i c$ the dielectric permeability ε of such a medium has the form.

$$\varepsilon = 1 - \frac{k_z^2}{k^2}\left[\frac{\omega_0^2}{\omega^2} - \frac{\Omega_0^2 \xi_n}{(\omega - n\Omega_i)^2}\right],$$ (3)

where $\omega_0 = \dfrac{4\pi e^2 n_0}{m_e}, \Omega_0^2 = \dfrac{4\pi e^2 n_0}{m_i}, \xi_n = I_n(\chi)e^{-\chi}, \chi = \dfrac{k_\perp^2 T_\perp}{m_i\Omega_0^2}$. Here k is the wave number,

k_z is the projection of k on the direction of the magnetic field. I_n is the Bessel function with imaginary argument index n, m_e and m_i are the electron and proton masses, respectively, n_0 is the plasma density and c is the light velocity. For such physical conditions in [28] is obtained a dispersion

$$\frac{\partial\varepsilon}{\partial\omega} = n_0\left[\frac{2a}{\omega^3} + \frac{2b}{(\omega - n\Omega_i)^3}\right],$$ (4)

where $a = \dfrac{4\pi e^2}{m_e}\dfrac{k_z^2}{k}, b = \dfrac{m_e}{m_i}a\xi$, which in the case when $\omega < n\Omega_i$ becomes anomalous

$\left(\dfrac{d\varepsilon}{d\omega} < 0\right)$. Then in the collisionless magnetized (H ~ 10^{-2}G) plasma of the outer

corona with an open magnetic field configuration, oscillations with periods T > 9.6×10^3 H ~ 0.1 s could carry negative energy in the case when the protons have anizotropic velocity distribution function.

4. Conclusion

The low excitation emission line registered episodically in the coronal spectrum pose a curios problem both in observational and theoretical aspects. This review paper illustrates the controversial state of the interpretations concerning the observational results and proposes some new theoretical ideas in the discussion on the possible coronal sources of these emission lines. However, further studies sustaning these ideas

138

are necessary to be carried out at forthcoming eclipses and I would recommend such studies to be included in the observational programmes for the 1999 total solar eclipse.

Acknowledgements The author wish to thank the organizers for the local and travel support. This work was partially supported by the Ministry of Education, Science and Technologies (Bulgaria) under grant F812/96.

References

1. Aly, M. K. (1955) Preliminary note on measures of coronal emission lines observed at the total solar eclipse, February 25, 1952, by B. Lyot and M. K. Aly, *Astrophys. J.*, **122**, 438-444.
2. Bappu, K. K. V., Bhattacharyya, J. C., and Sivaraman K. R. (1972) On emission lines of hydrogen, helium and ionized calcium seen on a coronal spectrogram of March 7, 1970 eclipse, *Solar Phys.* **26**, 366-369.
3. Grotrian, W. (1934), *Zs. f Astrophys.* **B8**, 129.
4. Deutsch, A. and Righini, G. (1964) An airborne observation of the coronal spectrum at the eclipse of July 20, 1963, *Astrophys. J.* **140**, 313-318.
5. Calacevich, A. (1953) Resultati prelliminari relativi allo spettro della corona esterna ottenuti nell'eclisi totale di sole del 25 febbraio 1952, *Convegno di Scienze Fisishe matematiche e Naturali, 14-19 settembre, 1952*, 186-192.
6. Migeotte, M. and Rosen, B. (1955), L' observation de l'eclipse solaire du 30 juin 1954, *Ciel Terre* **71**, 288-298
7. Parijskij, N. N. and Petrova, K. I. (1958) *Spectrometria Koronalnyh i chromosfernyh linij vo vremia zatmenia 25 febralia 1952, Polnye Solnetchnye zatmenya 25.II.1952 i 30.VI.1954*, Izd. Akad. Nauk USSR, 258-290.
8. Caccin, B., Moschi, G., Rigutti, M. and Falciani, R. (1971) Spectrometric analysis of the solar corona during the 12 November 1966 total solar eclipse, *Solar Phys.* **17**, 89-96.
9. Stellmacher, G. and Koutchmy, S. (1974) Study of low dispersion eclipse spectra: Observation of weak low excitation emission lines in the corona, *Astron. and Astrophys.* **35**, 43-48.
10. Gurtovenko, E. A. and Alikayeva, K. V. (1971) The physical conditions in inner corona derived from spectral data of the solar eclipse on 7 march, 1970, *Solar Phys.* **21**, 325-331.
11. Alikayeva K. V. (1975) On a cold emission in the solar corona, *Solar Phys.* **41**, 89-95.
12. Cavallini, F. and Righini, A. (1975) H and K (Ca II) emissions as observed in coronal spectrum in the July 20, 1963 solar eclipse, *Solar Phys.* **45**, 291-299.
13. Conway, J. G., Morris, W. F. and Andrews, C. F. (1967) Airborne Spectrographic Observations of the Solar Eclipse of November 12, 1966, *Astrophys. J.* **150**, 299-301.
14. Evans, C. D. (1970) A search for the K-line of Ca II during the eclipse of 7 March, 1970, *Solar Phys.* **14**, 157-158.
15. Kononovich, E. V. Smirnova, O. B. and Schukin, A. E. (1994) On possible CaII emission in the solar corona, in V. Rušin, P. Heinzel and J.-C. Vial (eds.) IAU Colloq. 144 "*Solar Coronal Structures*", VEDA PC, Bratsilava, pp. 589-591.
16. Öhman, Y. (1953) On some phenomena related to dark and bright prominences, *Convegno di Scienze Fisishe matematiche e Naturali, 14-19 settembre*, 1952, 156-163.
17. Leroy, Y. (1973) Emissions "froides" dans la coronne solaire, *Solar Phys.* **25**, 413-417.
18. Öhman, Y. (1973) Some comments on the low intensity Hα emission observed by J.-L. Leroy in the solar corona, *Solar Phys.* **28**, 399-402.
19. Dermendjiev, V. N., Mouradian, Z., Leroy, J.-L., Duchlev, P. I., (1994) Faint Hα Emission in the corona and its relation to the prominences, in V. Rušin, P. Heinzel and J.-C. Vial (eds.) IAU Colloq. 144 "*Solar Coronal Structures*", VEDA PC, Bratislava, 339-342.
20. Bouchard, O., Koutchmy, S., November, L., Vial, J.-C. and Zirker, J. B. (1994) Very high resolution analysis of the dynamics of a coronal plasmoid, in V. Rušin, P. Heinzel and J.-C. Vial (eds.) IAU Colloq. 144 "*Solar Coronal Structures*", VEDA PC, Bratislava, 593-596.
21. Dermendjiev, V. N., Mouradian, Z., Duchlev, P. I., Leroy, J.-L. (1994) Faint Hα Emission in the corona: Morphological, Situational and Hydrodynamic Analysis", *Solar Phys.* **149**, 267-277.
22. Gnevyshev, M. N. and Gnevysheva, R. S. (1963) Helium emission in the solar corona, in J. W. Evans (ed.) IAU Symp. 116 "The solar corona", Acad. press NY&London, 241.

23. Zirin, H., James, R. and Watson, D. K. (1964) Identification of the Spurious D3 helium emission from the solar corona, *Astron. J.* **69**, 565.
24. Gnevyshev M. N. and Gnevysheva, R. S. (1956) Novye dannye o solnechnoj korone, *Izv. Krimskoj Obs.* tom XVI, 212-215.
25. Shklovskij, I. S. (1962) Fizika Solnechnoj Corony, Fizmatgiz, Moskva, 205, (in russian).
26. Öhman, Y. (1958), On the absorption effects in prominences possibly connected with a magnetic cooling mechanism proposed by L. Marshall, *Astrophys. J.* **128**, 92-94
27. Marshall, L. (1957) Coronal streaming in solar sunspot prominences, *Astrophys. J.* **126**, 177-184.
28. Dermendjiev, V. N., Mouradian, Z., Leroy, J.-L. (1994), On the Problem of the "Emission Froid" in the Solar Corona, *Comptes rendus de l'Acad. Bulgare des Sci.* **47**, No. 7, 5-7.
29. Sturrock, P. A. (1960) In what sense do slow wave carry negative energy?, *J. of Applied Phys.* **31**, 2052-2056.
30. Kodomtsev, B. B., Mikhailovsky, A. B. and Timofeyev, A. V. (1964) Negative energy waves in dispersive media, *Journal Eksperimentalnoi i teoretichnoi Fiziki* tom **47**, 2266 - 2268 (in russian).

CREATION OF PROMINENCES AND FILAMENTS
New Vista on Familiar Sunscapes

V. GAIZAUSKAS

Herzberg Institute of Astrophysics,
National Research Council of Canada
100 Sussex Drive, Ottawa K1A 0R6, Canada

Abstract. Because chromospheric filaments are invariably associated with polarity inversions in the photospheric magnetic field, they have long been used to trace the evolution of global patterns of magnetic flux. Individual quiescent filaments may last a single solar rotation or less, but it is commonly believed that the channels in which they form are much longer-lived. Attempts to trace the inverse process - the origin of a filament channel and a filament during the evolution of specific patches of flux - are of more recent origin. New observations are now forcing a revision in our ideas about filament formation. After reviewing some of the new key facts, a recent case study is reported in which a filament channel and filament are seen to form on the edge of a growing activity complex. The evidence points to the formation of this channel as a surface phenomenon driven by emerging magnetic flux. The filament forms when its channel is constricted between the expanding new region and a pre-existing plage.

1. Introduction

Solar filaments form in filament channels - voids without plages or chromospheric fine structure such as spicules or fibrils rooted in enhanced magnetic elements [9]. Channels overlie polarity inversions: they follow along the division between opposite polarities in the line-of-sight magnetic fields ($B_\parallel$) measured in the photosphere. The mere existence of a channel in the belt of active latitudes is necessary but insufficient to form a filament.

In off-band Ha images a channel appears near the center of the disk as a long, narrow feature in which dark fibrils to either side of the channel are aligned parallel to it [1]. Chromospheric "threads" and "fibrils" are distinguished from "filaments" as follows [1]: "threads" are thin, single structures with two well defined ends, each rooted in plages or network elements of opposite magnetic polarity; but "fibrils", such as those which border a filament channel, are open-ended structures with only one well-defined end rooted in a plage or plagette. Fibrils stream out of plagettes in antiparallel directions on opposite sides of a polarity inversion, leading Foukal [1] to conclude that the orientation of fibrils signifies a predominantly horizontal component for the magnetic field along, and not across, the axial direction of filament channels.

141

Z. Mouradian and M. Stavinschi (eds.),
Theoretical and Observational Problems Related to Solar Eclipses, 141-148.

Channels have a hemispheric asymmetry [8]: when quiescent filaments are viewed from the positive polarity side, a majority of channels in the northern hemisphere have fields directed to the right, while in the southern hemisphere the fields are directed toward the left. Thus channel fields have a definite orientation - either "dextral" or "sinistral" - which predominates in each hemisphere. Channels with a common orientation can join to form very long and persistent structures [8].

Until now, no one has described the formation of a filament channel. Although filaments and channels abound at existing or decaying active regions, an extensive search [3] of the film archive of the Ottawa River Solar Observatory (ORSO) failed to catch a channel in the act of formation where magnetic flux density was stable or declining. This is surprising because previous studies [4, 11] suggest that filaments form predominantly in areas of decaying magnetic flux at the borders of, or between, activity complexes. Instead, as summarized and discussed below, the search uncovered the birth of a channel and the subsequent development in it of a filament where a new activity complex was emerging in a previously quiet, near-equatorial, area.

2. Observations on the Birth of a Filament Channel and a Filament

The following list summarizes the essential observations of McMath 16166 on 19-25 July, 1979, extracted from the detailed description by Gaizauskas *et al.* [3]:

1. Flux emerging in adjacent bipoles (inside the oval, Figure 1, top panel) creates a compact multipolar environment which prevails from within a day after the first Arch Filament System (AFS) appears up to the formation of a filament several days later.
2. Ha fibrils organize as a distinctive pattern of parallel structures in the trailing sector (at B, Figure 1) of the small multipolar activity complex. This phase of fibril alignment occurs within a few hours; it coincides with rapid sunspot growth (note the strengthened AFS in Figure 1, lower panel).
3. The growing activity complex opens a 'moat' between the new growing sunspots and the background of quiet network fields, especially at B (lower panel, Figure 2).
4. The organized pattern of fibrils at B lies on a polarity inversion between the trailing (+ve) polarity of the expanding new activity complex and the leading (-ve) polarity flux of the remnant plage of an old region (Figure 2). Oppositely-oriented streaming fibrils to either side of the polarity inversion at B confirm that their pattern defines a channel and that it is *sinistral* .
5. The organized pattern of fibrils does not coalesce in any systematic way into an Ha filament. Structures with the dimensions of a filament often grow inside the organized pattern, but they are short-lived (<< 1 day).
6. A stable filament forms three days after the channel is created, with one end passing through a point of tight convergence between flux dispersing from the new, rapidly expanding activity complex and flux redistributing stochastically in an old plage (at F_1 in both panels of Figure 3).

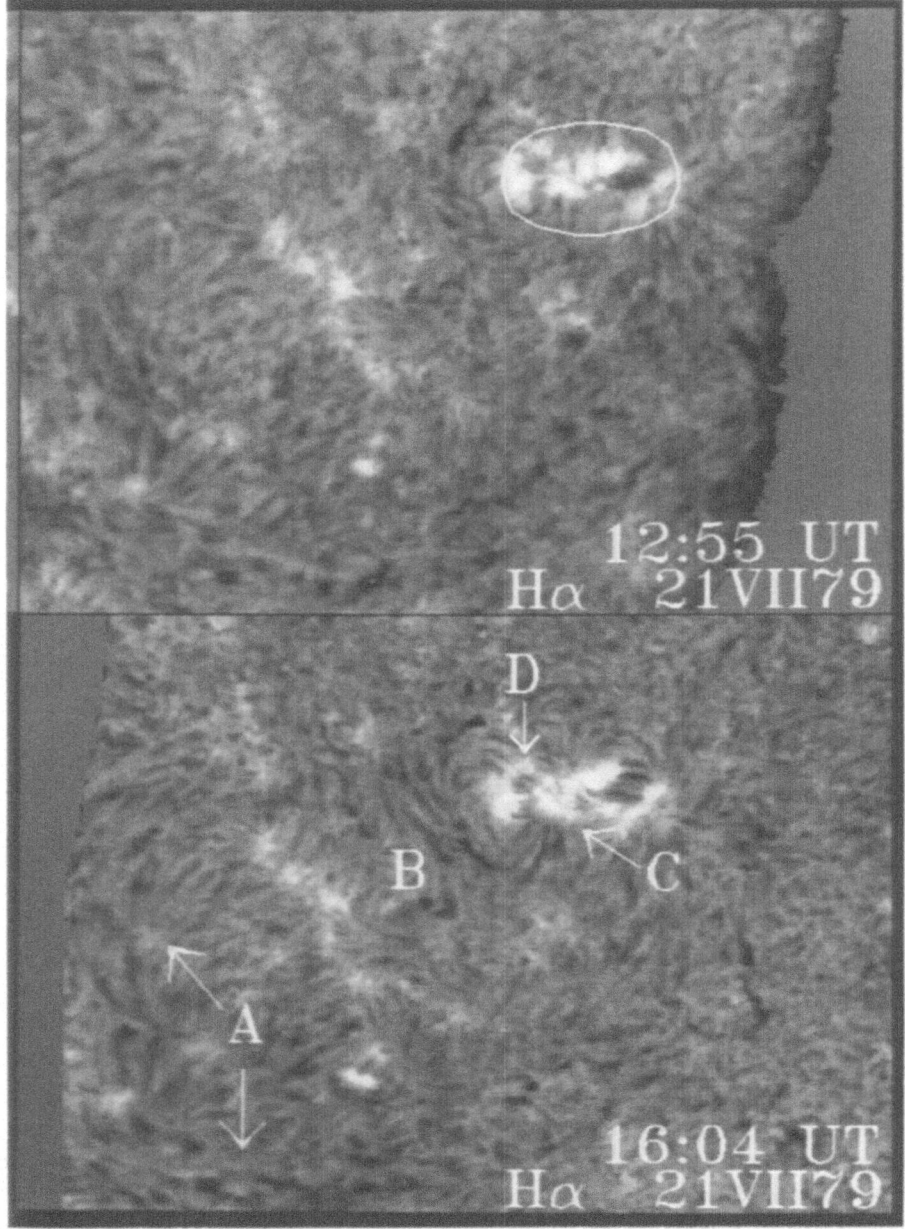

Figure 1. Rapid growth of McMath 16166 (oval outline) on 21 July, 1979. Note the strengthened alignment of fibrils (in 3 hr) ringing the eastward (left) side of the region and of the Arch Filament System at the western (right) end of the oval. The two filament channels discussed in the text are at A (marked with arrows, lower right) and B (with terminations near C and D). Each frame is 455"x324" and is oriented N at top, W right. Images are composites of four adjacent fields of view from the ORSO film archive.

144

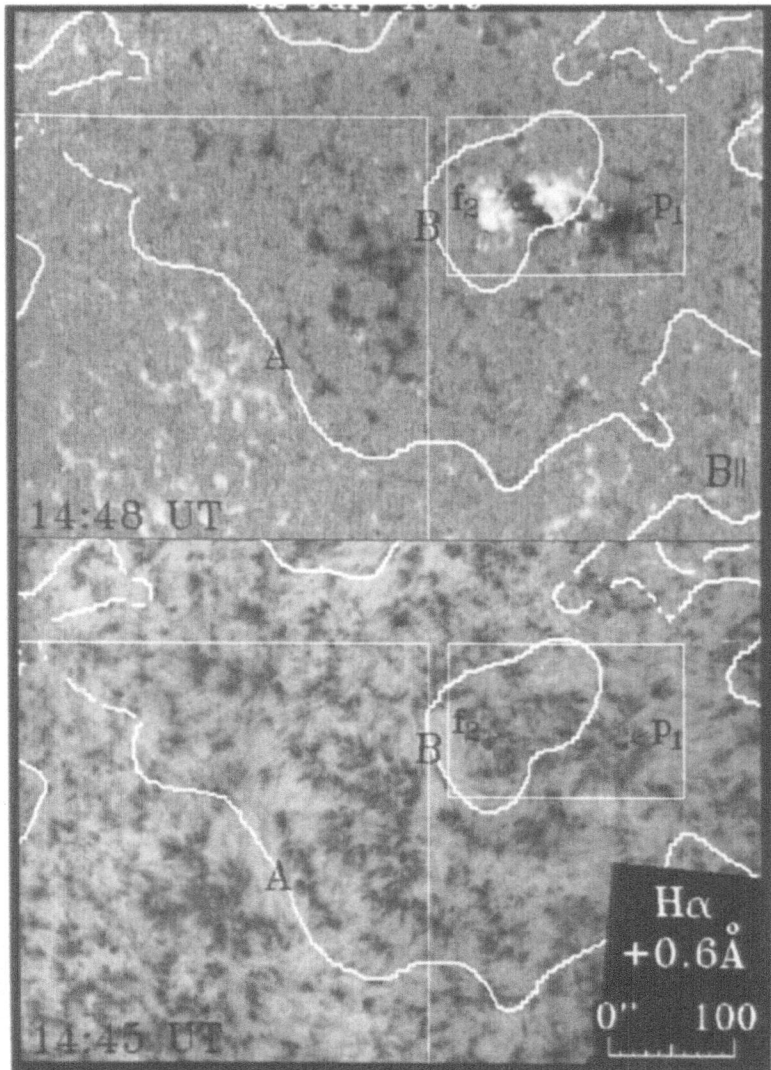

Figure 2. Polarity inversion lines overplotted as white contours on magnetogram (NSO/Kitt Peak, top) and off-band Hα filtergram (ORSO, bottom) for the same field of view on 22 July, 1979. The smaller box encloses newly emerged activity complex McMath 16166; the larger box outlines most of decaying region McMath 16159. A and B mark locations of filament channels discussed in the text. p_1 marks the first leading polarity (black = -ve) spot, f_2 the largest trailing polarity (white = +ve) spot in the 2-bipole activity complex. Orientation: N at top, W right.

7. The new filament is morphologically entirely different from preceding dark, long fibrils; although small, it has the barbed edges and bright rim characteristic of quiescent prominences.

8. The filament survives up to its W limb passage (3 days later), but not until the next solar rotation; filamentary strands then appear in other parts of the same filament channel.

9. Other polarity inversions, where filaments do not form, either have no channel (e.g. at C and D in Figures 1 and 2 or at F_2 in Figure 3) because they are covered by transverse fields or because they lack points in the channel with strongly converging magnetic flux of opposite polarities (e.g., the channel at A, Figures 1 and 2).

3. Discussion

The simplest explanation for the formation of the channel at B (Figures 1 and 2) is that it is due to the extended magnetic field of the growing activity complex. The magnetic field in the channel cannot, however, be a potential one. Hα fibrils there do not radiate out of the trailing positive pole of McMath 16166 before closing back, north and south, to its leading negative polarity partner as they would for a potential field (e.g. in the vicinity of p_1 at the leading end of the activity complex (Figure 2)). We see instead a swirled arc of fibrils at B which is consistent with a pattern derived by Low [5]. His magnetostatic equilibrium model creates a similar pattern of magnetic field lines in the lower atmosphere above a simple bipolar active region when he includes both field-aligned and cross-field currents. In this more general case a Lorentz force due to cross-field currents, in balance with plasma pressure and gravity, modifies the magnetic topology low in the atmosphere but leaves it force-free at higher levels. We may conjecture that a similar combination of cross-field and field aligned currents, generated during the growth of the compact activity complex, contributes to the formation of a system of flat horizontal loops around the periphery of McMath 16166.

We picture the flat loops as a 'seed' field for the filament channel that develops as McMath 16166 expands. The presence of such low horizontal fields enhances the possibility of reconnections with small-scale network fields, thereby maintaining and reconfiguring the horizontal field. This is consistent with scenarios developed empirically [8] and theoretically [12]. Additional observations and modeling of non-potential fields are required before we are able to understand it.

Given these new data, Mackay *et al.* [6] attempted, unsuccessfully, to model the field lines for the channel at B with a potential field arising from multiple magnetic sources distributed on both sides of the channel to represent the measured B∥ flux. But a more advanced model [6], a force-free approximation to the field of this activity complex, produces low-lying field lines with a dominant horizontal component which agrees well with the observed patterns of Hα fibrils at B in Figure 1. Once the filament forms, local potential models of the channel suffice, provided a strong horizontal field component is superposed to obtain the correct connectivity for the path of the filament in the channel.

Converging fields are evidently necessary to form filaments; their link with canceling flux has already been noted [7] as commonplace for a large sample of

146

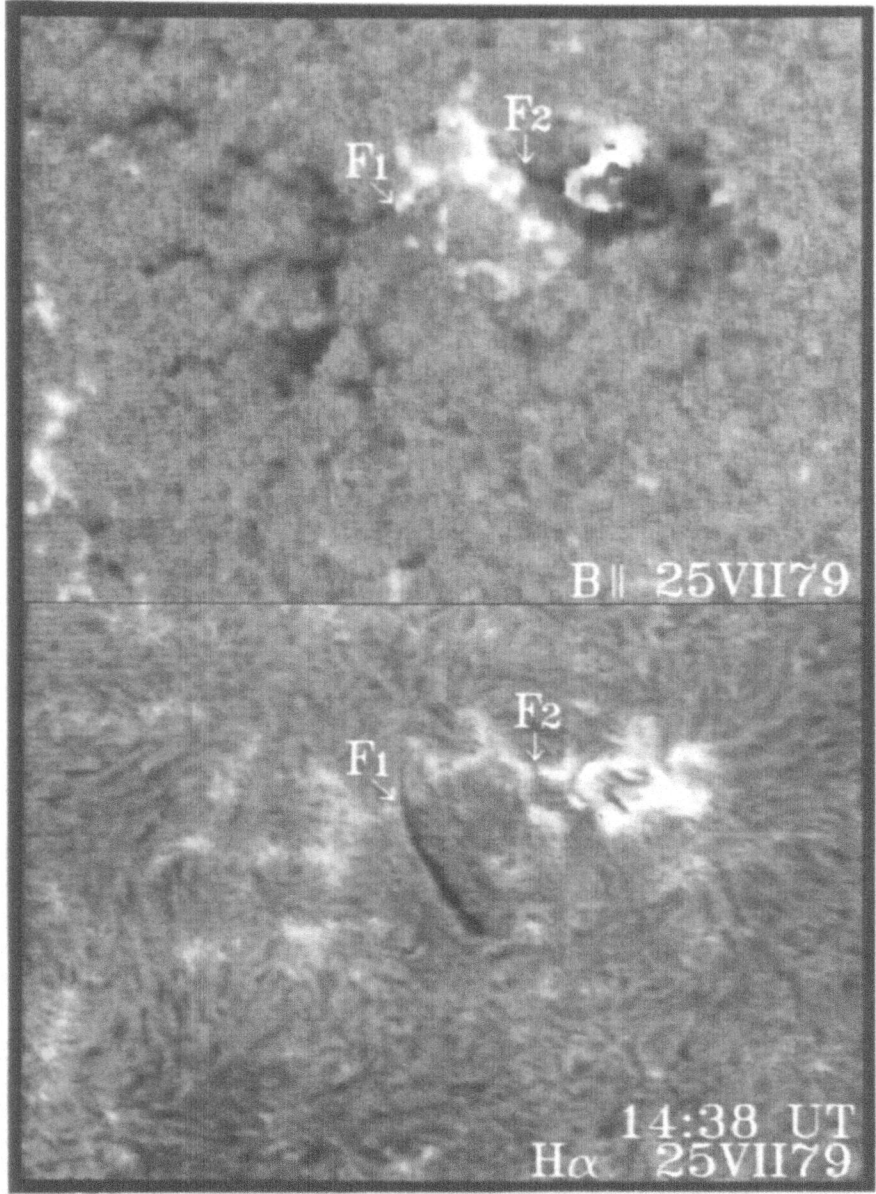

Figure 3. A stable filament forms in McMath 16166 on 25 July, 1979 at a point where +ve (white) and -ve (black) magnetic polarities come into contact (at F_1, in both panels). No filament forms at another point of convergence (F_2) because the horizontal field there is across, not along, the polarity inversion as indicated by Hα fibrils. Each frame is 455"x324" with N at top, W right. Upper panel: NSO/Kitt Peak magnetogram. Lower panel: composite filtergram of four adjacent fields of view from the ORSO film archive.

filaments. What possible significance can converging regions have to filament formation? One effect is to form an arcade over a channel in order to anchor its filament. Another possibility is that a current sheet forms between the merging flux patterns, thereby pumping mass along an already formed channel with its dominant horizontal field component. By way of contrast, the large channel at A (Figures 1 and 2), which does not develop a filament, also does not narrow significantly during this passage across the solar disk. Clearly, more information is needed about the detailed magnetic structure of filaments.

4. Summary

A filament channel has been captured for the first time in the act of formation. It built up during the emergence of a new activity complex and its subsequent interaction with the existing flux of an adjacent decaying active region. The channel neither preceded nor coincided with the initial arrival of new flux tubes. Similarly, a filament formed in the channel by a dynamic interaction at the surface of already emerged magnetic elements, not by buoying up a prefabricated flux tube as proposed by Rust and Kumar [10].

The quintessential characteristic of this filament channel is a horizontal magnetic field directed along the length of the channel, in accord with earlier observations [1, 7]. In this case, however, the pattern of fibrils in the new channel evolves too rapidly and is too well organized to be ascribed entirely to reconnections among many small magnetic elements converging collectively toward a polarity inversion line [7, 9, 12]. The timing of events - improved alignment of Ha fibrils concurrent with a fresh injection of magnetic flux - points to the distortion of the magnetic topology of the activity complex into a non-potential form by electric currents generated by the newly emerging flux.

A convergence (and probably cancellation) between opposite magnetic polarities in a filament channel crystallizes the growth of this filament. The importance of flux convergence has already been stressed in earlier studies [7, 8]. Here the converse is also demonstrated: flux convergence at a polarity inversion does not create a filament when no filament channel is present; and a filament will not form in a filament channel between plages of opposite polarity if they do not do not converge.

5. Recommendation

The speculation presented above, namely, that convergence of flux at a polarity inversion produces a localized current sheet which then directs mass along the horizontal field in a filament channel to create the visible filament, has a verifiable consequence. If correct, there should be hot, dense plasma nearly in contact with cooler plasma in the spine of quiescent prominences of the type described in this study: filaments lying between adjacent activity complexes. Although invisible in Ha such plasma could, in principle, be detected: as a separate structure in the UV or XUV from spacecraft, or as done in the 11 July, 1991 eclipse, as a mismatch in shapes between white-light and Ha images of limb prominences taken during a total eclipse with high spatial resolution [2].

Thus for the 1999 total eclipse filaments should be observed at high spatial and temporal resolution prior to their arrival at the limb on the day of the eclipse in order to pick appropriate prominences as targets. The aim would be to test the above hypothesis, i.e. that hot, dense plasma may exist above a filament, possibly within the filament cavity.

Acknowledgments

I am indebted to the LOC of this NATO Advanced Research Workshop for their generous hospitality during my stay in Bucharest and Sinaia, and to Jack Zirker and Duncan Mackay for discussions prior to the presentation of this paper.

References

1. Foukal, P. (1971) Morphological Relationships in the Chromospheric Ha Fine Structure, *Solar Phys.*, **19**, 59-71.

2. Gaizauskas, V., De Luca E., Golub, L, Jones, H.P., and November, L. (1996), Observations of a Quiescent Prominence Straddling the Solar Limb during the Total Eclipse of 11 July 1991, in Y. Uchida, T. Kosugi, and H. S. Hudson (eds.), *Magnetodynamic Phenomena in the Solar Atmosphere,* IAU Colloq. **153**, Kluwer Academic Publishers, Dordrecht, in press.

3. Gaizauskas, V., Zirker, J.B., Sweetland, C., and Kovacs, A. (1996) Formation of a Solar Filament Channel, *Astrophys. J.* (in press).

4. Kiepenheuer, K.O. (1953) Solar Activity, in G.P. Kuiper (ed.) *The Sun,* Chicago University Press, **322**.

5. Low, B.C. (1992) Three-Dimensional Structures of Magnetostatic Atmospheres. IV. Magnetic Structures Over a Solar Active Region, *Astrophys. J.*, **399**, 300-312.

6. Mackay, D.H., Gaizauskas, V., Rickard G., and Priest, E.R. (1996) Force-Free and Potential Models of a Filament Channel in which a Filament Forms, *Astrophys. J.*, submitted.

7. Martin, S.F. (1990) Conditions for the Formation of Prominences as Inferred from Optical Observations, in V. Ruzdjak and E. Tandberg-Hannsen (eds.), *Dynamics of Quiescent Prominences*, Lecture Notes in Physics **363**, Springer-Verlag, Berlin, 1-44.

8. Martin, S.F., Bilimoria, R., and Tracadas, P.W. (1994) Magnetic Field Configurations Basic to Filament Channels and Filaments, in R.J. Rutten and C.J. Schrijver (eds.), *Solar Surface Magnetism*, NATO ASI Series C-433, Kluwer Academic Publishers, Dordrecht, 303-338.

9. Martres, M.-J., Michard, R., and Soru-Iscovici, I. (1966) Etude Morphologique de la Structure Magnétique des Régions Actives en Relation avec les Phénomenes Chromosphériques et les Erptions Solaires. II Localisation des plages brillantes, filaments et éruptions, *Ann. Astrophys.*, **29**, 249-253.

10. Rust, D.M. & Kumar, A. (1994) Helical Magnetic Fields in Filaments, *Solar Phys.* **155**, 69-97.

11. Tang, F. (1987) Quiescent Prominences - Where do they Form? *Solar Phys.*, **107**, 233-237.

12. van Ballegooijen, A.A. and Martens, P.C.H. (1989) Formation and Eruption of Solar Prominences, *Astrophys. J.*, **343**, 971-984.

UNSOLVED PROBLEMS IN PROMINENCE RESEARCH

T.G. FORBES
Institute for the Study of Earth, Oceans, and Space
University of New Hampshire, Durham, NH 03824, USA

ABSTRACT. Understanding the magnetic field structure of prominences is a challenge to both observers and theorists. A model for the magnetic field of a prominence must be able to fulfill several basic functions. The model must explain why prominences erupt and prescribe the conditions that lead to eruption. Additionally, the pre-eruption field configuration must be consistent with the observed structure of quiescent prominences and account for their gravitational and thermal stability. So far, there are no models which can satisfy all these conditions.

1. Introduction

Although there are two-dimensional prominence models which provide plausible mechanisms for the eruption of a prominence [9, 20], these models do not satisfactorily describe structural features such as the prominence legs (*i.e.* barbs). On the other hand, prominence models which use potential fields to explain the overall three-dimensional structure [14] provide no mechanism to explain why prominences erupt. Progress has been slow in developing a completely satisfactory model because of the basic conundrum posed by the magnetic field in which the prominence lies. Observationally, the field is difficult to measure, while theoretically, its is difficult to solve the MHD equations which govern the behavior of the field in three dimensions.

Difficulties in reconciling the various models suggest that the magnetic forces which drive an eruption arise in the large scale field structures which surround a prominence rather than within the prominence itself. The large scale field is thought to contain field aligned currents which exist independently of the prominence. Knowing the location and evolution of these currents is especially important for understanding the nature of the eruptive mechanism of prominences, but these currents are even more difficult to detect than the magnetic field.

149

Z. Mouradian and M. Stavinschi (eds.),
Theoretical and Observational Problems Related to Solar Eclipses, 149-157.
©1997 *Kluwer Academic Publishers.*

150

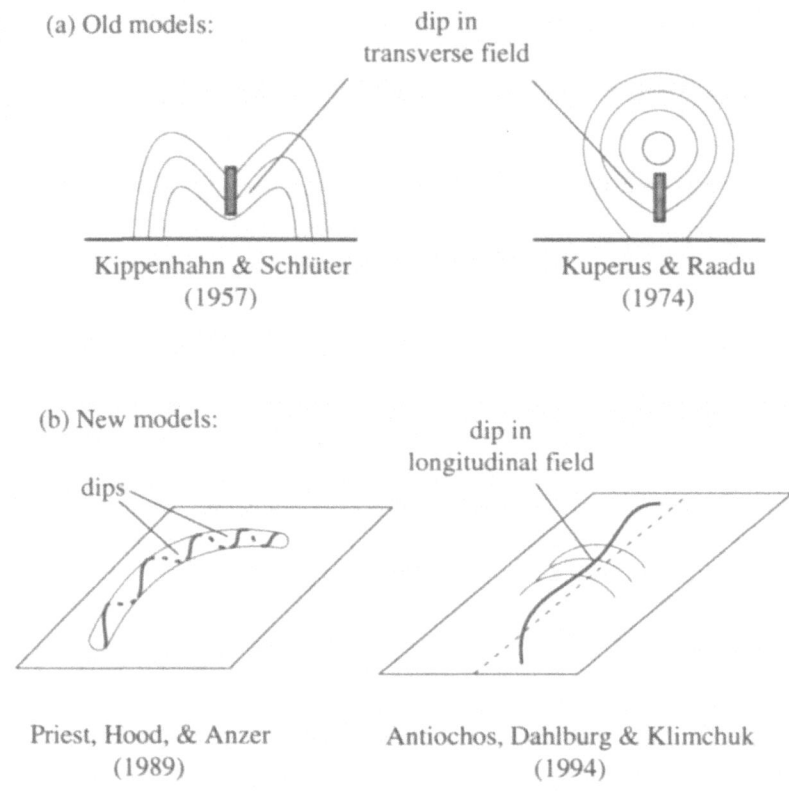

(a) Old models: dip in
transverse field

Kippenhahn & Schlüter   Kuperus & Raadu
(1957)   (1974)

(b) New models:

dip in
longitudinal field

dips

Priest, Hood, & Anzer   Antiochos, Dahlburg & Klimchuk
(1989)   (1994)

Figure 1. Various well known models for supporting prominences

2. Gravitational Support and Thermal Stability

Following earlier work by Menzel [32] and Dungey [5], Kippenhahn and Schlüter [10] constructed what is still probably the best know model of a prominence. This model assumes the configuration shown in Figure 1a and gives a description of the field and plasma only within the vicinity of the prominence itself. Kippenhahn and Schlüter's model is not really a global model since it says nothing about how the field lines are anchored to the photosphere. Their model also ignores the very strong field component that exists perpendicular to the plane of the figure.

In order to support the weight of the prominence against gravity, the Kippenhahn and Schlüter model requires that the mass of the prominence be sufficiently concentrated to form a dip at the top of a magnetic loop [2]. The radius of curvature of the dip R , can be expressed in terms of the gravitational scale height and plasma b of the prominence. Balancing the upward magnetic force due to the dip against the downward pull of gravity gives

$$j B / c = r g \qquad (1)$$

where j is the current density created by weight of the prominence, B is the ambient magnetic field, c is the speed of light, r is the prominence density, and g is the gravitational acceleration. The current density, j, can be expressed very roughly in terms of the radius of curvature, R, as

$$j = cB / (8pR) \tag{2}$$

which assumes that the perturbation field due to the dip is of the same order as B. Combining (1) and (2) gives

$$R = Hg/b \tag{3}$$

where $H_g = P/(rg)$ is the prominence's gravitational scale height, P is pressure, and $b = 8pP/B^2$ is the plasma beta parameter inside the prominence. Observations and

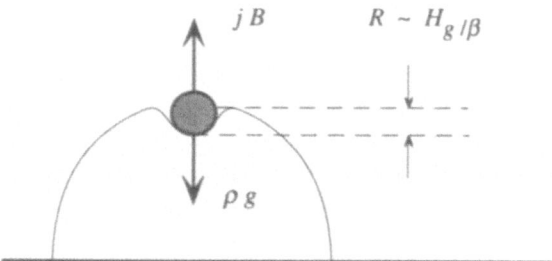

Figure 2 *Diagram for estimating the radius of curvature* R *,of the dip created by the prominence mass in Kippenhahn-Schlüter model.*

potential extrapolation of the photospheric magnetic field imply that b is in the range of 0.1 to 0.001 which, according to Equation (3), requires R to be 10 to 1000 times greater than H_g. Since the observed thickness of prominences is on the order of their gravitational scale height, the radius of curvature of the dip produced by the weight of the prominence is much larger than the prominence's thickness. In other words, the observed concentration of mass in the prominence is not sufficient to form a localized dip of the same size as the thickness of the prominence. This result implies that the unperturbed field in the region where a prominence forms must be curved upwards or at least be very flat.

The difficulty of forming a prominence and maintaining it stability in the Kippenhahn and Schlüter model led [11] to propose the inverse polarity configuration shown in Figure 1a. After the discovery that most quiescent region prominences do in fact have an inverse polarity [12] interest in the Kippenhahn and Schlüter model declined.

(a)

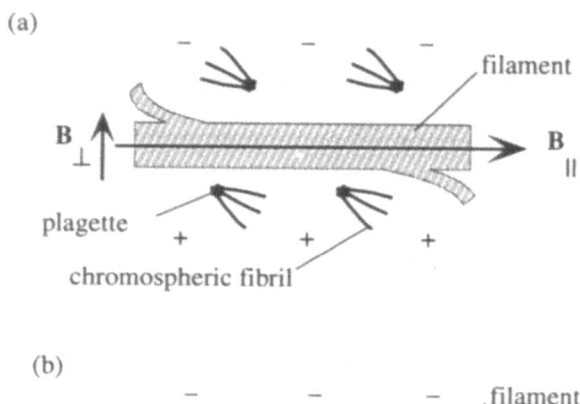

(b)

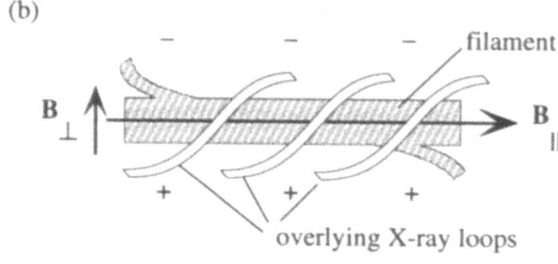

Figure 3. Morphological features of prominences in (a) Hα and (b) X-rays. The magnetic field, B∧, is the transverse component of the potential field due to photospheric sources.

Subsequent models [3, 4, 13, 15, 23] have tended to be of the Kuperus-Raadu type.

Neither the Kippenhahn-Schlüter nor the Kuperus-Raadu models address the thermodynamics of prominences, nor do they address the global aspects of gravitational support and mechanical stability of the overall configuration. Recently, Antiochos, Dahlburg, and Klimchuk [3] have developed a three-dimensional model which creates a dip in the longitudinal field, instead of the transverse field, as shown in Figure 1b. An overlying transverse field produces a dip in the longitudinal field by pulling down on it, and since the configuration is also sheared, the transverse field has an inverse orientation in the region of the dip. Another recent model which considers both transverse and longitudinal field components is the flux-rope model shown at the right of Figure 1b [24].

Sometimes prominences disappear simply by fading in place without any obvious eruption or change in their shape, and after the prominence has faded, high temperature plasma is seen in the same region previously occupied by the prominence [22, 28]. Because the temperature of this plasma is slightly higher than that of the ambient corona, it is most easily observed in extreme ultra-violet emission lines as a kind of ghost image of the original prominence. This kind of disappearance is thought to result from a heating of the prominence by an external agent such as a nearby flare. The prominence can also reappear and disappear several times in the same magnetic channel, and only part of the prominence may disappear while the rest remains.

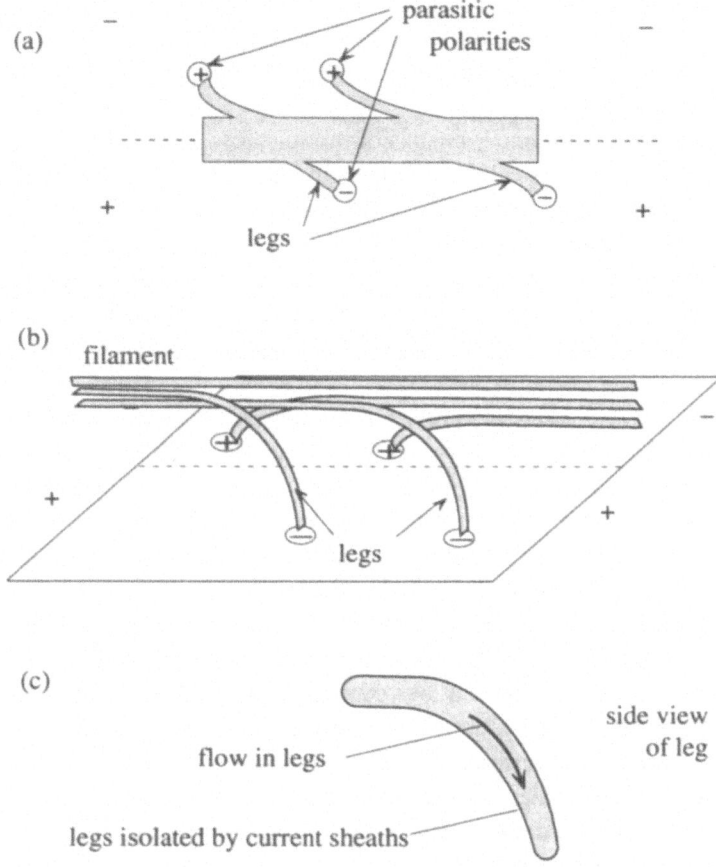

Figure 4. A model by Martin [17] to explain some of the morphological features of inverse prominences.

3. Eruption Mechanism and Morphology

Nearly all models of erupting prominences are based on the principle that the energy which drives the eruption comes from magnetic energy associated with coronal currents[31]. The currents may be transported there by the emergence of a flux rope from the convection zone or by moving the footpoints of an arcade of magnetic loops so as to develop a sheared field.

When a prominence erupts, magnetic field lines mapping from the ejected plasma to the photosphere are stretched outwards to form an extended, open field structure. The opening of the field creates an apparent paradox since the stretching of the field lines implies that the magnetic energy of the system is increasing whereas storage models require it to decrease. Aly [1] and Sturrock [30] have shown that for simply connected fields, the fully-opened state is in fact the highest energy state of the system. Therefore, to create an eruption with a storage model, one must either start with a field which is not simply connected to the photosphere or avoid creating a field which is everywhere open.

As previously mentioned, models of prominences must not only explain how an eruption occurs, they must also explain their morphological features. Several of these features are illustrated in Figure 3. When viewed from above, prominences (or filaments as they are usually referred to when seen from above) appear to be made up of individual threads which suggest a sheared field. The chirality of the shear is also reflected in the prominence legs (or barbs) which extend outward from main body of the prominence towards the photosphere [18]. In the chromosphere below the prominence, chromospheric structures known as plagettes also exhibit they chirality as the prominence itself. A plagette consists of a group of fibrils which radiate outward from a center. The orientation of the fibrils within the plagette is the same as the threads within the prominence, and the fibrils radiate outwards, away from the axis of the prominence. By contrast, overlying X-ray loops (as observed by the Soft X-ray Telescope on the Yohkoh satellite) appear to have a chirality which is opposite to that observed in the filament itself as illustrated in Figure 3b [19].

Despite the fact that the prominence legs are conspicuous features, most models tend to ignore them. Figure 4 shows a relatively recent model by Martin [17] which proposes that the prominence legs correspond to field lines mapping from the body of the prominence to small regions of parasitic (or reverse) polarity on either side of the magnetic polarity inversion line in the photosphere (Figure 4a and Figure 4b). Because the parasitic polarities are located at the center of the large-scale convection cells, the legs are predicted to terminate in the center of these cells. The model successfully accounts for the inverse polarity of the field within the prominence relative to the photosphere.

Martin's model requires the legs to be field aligned structures in which dense, cool plasma is not statically supported. Consequently, the model predicts that there should be flow in the legs, but this flow could be either downwards or upwards depending on whether there is some dynamic process associated with reconnection at the parasitic polarities. Although steady flows are observed in prominences [29], it is not yet clear if these flows are consistent with flows in the legs [15].

Another special feature of the Martin model is that the magnetic flux in the leg must be surrounded by a current sheath in order to isolate the parasitic polarity field from the surrounding field (Figure 4c). Otherwise, field lines from the parasitic polarity regions connect with the surrounding field lines and do not reach the height of the prominence. As originally envisioned, the model neither accounts for the outward orientation of the plagettes nor the reverse chirality of the overlying X-ray loops. Nor does the model include the coronal currents which are required to explain an eruption.

(a) View from side

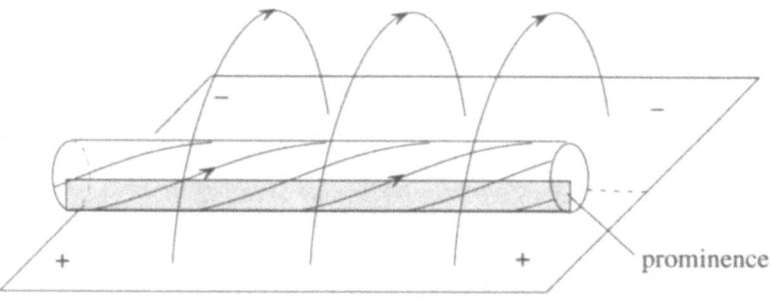

prominence

(b) View from above

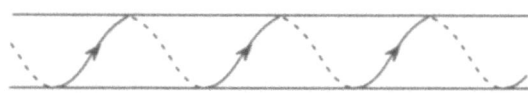

Figure 5. The flux-rope model for inverse prominences [25].

One prominence model which can account for an eruption, but only within the context of two-dimensional MHD, is the flux rope model [7, 9, 21, 33, 34]. A schematic diagram of this model is shown in Figure 5. The prominence is located in the bottom half of a helical flux rope where the field is of inverse polarity with respect to the photosphere, and the current which flows along the flux rope provides the energy for the eruption of the field [8]. The outward orientation of the plagettes is explained by the fact that the field in their vicinity tilts out and around the flux rope [6]. The apparent reverse chirality of the overlying X-ray loops reflects the orientation of the field lines in the top half of the flux rope [26, 27].

Although the flux-rope configuration in Figure 5 accounts for several morphological features, it does not explain why legs exist. One possible way to remove this limitation is to add the parasitic polarities with Martin's model to the flux-rope model as shown in Figure 6. Adding the parasitic polarity connects some of the field lines within the flux rope to the parasitic polarity, and makes it easier for the field lines in the parasitic polarity to reach the altitude of the prominence.

4. Summary

At the present time the geometry of the magnetic field in and around prominences remains unclear. A few tantalizing facts about the field are known such as its normal field component at the photosphere and its average strength and orientation inside the

156

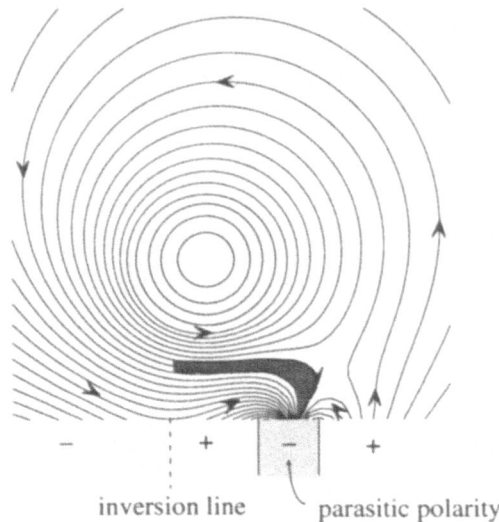

inversion line   parasitic polarity

Figure 6. A diagram illustrating how a leg (barb) might be formed in the flux-rope model by adding a parasitic polarity.

prominence. From the dynamics of prominence eruptions we can also infer that field contains a substantial current which is probably field aligned. Models have been proposed to explain various aspects of prominences such as how they are supported against gravity, why they erupt, and why they have the form that is observed. So far, no model exists which can explain all these aspects.

A feature of prominences which is particularly puzzling is the leg (or barb). Observationally, it is still an open question whether it is a static structure supported against gravity by a horizontal magnetic field or a dynamic structure which is field aligned. Developing new models, or extending existing ones, to explain the existence of legs would help improve our understanding of the magnetic field configuration in which prominences are imbedded.

The work was supported by NATO and by NASA grants NAGW-3463 and NAG5-1479 to the University of New Hampshire.

References

1 Aly, J.J.: (1985), *Astron. Astrophys.*, **143**, 19.

2 Amari, T., Hood, A.W., Priest, E.R., Démoulin, P., and Browning, P.K.: (1991), *Astron. and Astrophys.*, **241**, 604.

3 Antiochos, S.K., Dahlburg, R.B., and Klimchuk, J.A.: (1994), *Astrophys. J.*, **420**, L41.

4 Démoulin, P., and Forbes, T.G.: (1992), *Astrophys. J.*, **387**, 394.

5 Dungey, J.W.: (1953), *Monthly Notices Roy. Astron. Soc.*, **113**, 180.

6 Filippov, B.P.: (1995), *Astron. Astrophys.*, **303**, 242.

7 Forbes, T.G.: (1990), *J. Geophys. Res.*, **95**, 11919.

8   Forbes, T.G.: (1992), in *Eruptive Solar Flares*, Svestka, Z., B.V. Jackson and M.E. Machado (eds.), Springer-Verlag, New York, p. 79.

9   Forbes, T.G., and Priest, E.R.: (1995), *Astrophys. J.*, **446**, 377.

10 Kippenhahn, R., and Schlüter, A.: (1957), *Zs. Astrophys.*, **43**, 36.

11 Kuperus, M., and Raadu, M.A.: (1974), *Astron. Astrophys.*, **21**, 189.

12 Leroy, J.L., Bommier, V., and Sahal-Brechot, S.: (1983), *Solar Phys.*, **83**, 135.

13 Low, B.C., and Hundhausen, J.R.: (1995), *Astrophys. J.*, **443**, 818.

14 Mackay, D.H., and Priest, E.R.: (1996), *Solar Phys.*, **167**, 281.

15 Malherbe, J.M., and Priest, E.R.: (1983), *Astron. Astrophys.*, **123**, 80.

16 Malherbe, J.M., Schmieder, B., Ribes, E., and Mein, P.: (1983), *Astron. Astrophys.*, **119**, 197.

17 Martin, S.F.: (1994), *Bull. American Astron. Soc*, **26**, 185.

18 Martin, S.F., Bilimoria, R., and Tracadas, P.W.: (1994), in *Solar Surface Magnetism*, Rutten, R.J. and C.J. Schrijver (eds.), Kluwer, Dordrecht, p. 303.

19 Martin, S.F., and McAllister, A.: (1996), in *Proceedings of the Chapman Conference on Coronal Mass Ejections held at Bozeman, Montana*, Crooker, N.U. and J. Joslyn (eds.), AGU, Washington, D.C., p. submitted.

20 Mikic, Z., and Linker, J.A.: (1994), *Astrophys. J.*, **430**, 898.

21 Molodenskii, M.M., and Filippov, B.P.: (1987), *Soviet Astron. (English. Translation)*, **31**, 564.

22 Mouradian, Z., Martres, M.J., and Soru-Escaut, I.: (1986), in *Coronal and Prominence Plasmas*, Poland, A.I. (ed.) NASA, CP-2442, p. 221.

23 Pneuman, G.W.: (1983), *Solar Phys.*, **88**, 219.

24 Priest, E.R., Hood, A.W., and Anzer, U.: (1989), *Astrophys. J.*, **344**, 1010.

25 Priest, E.R., van Ballegooijen, A.A., and Mackay, D.H.: (1996), *Astrophys. J.*, **460**, 530.

26 Rust, D.M., and Kumar, A.: (1994), *Solar Phys.*, **155**, 69.

27 Rust, D.M., and Kumar, A.: (1996), *Astrophys. j.*, **464**, L199.

28 Schmahl, E.J., Kundu, M.R., Strong, K.T., Bentley, R.D., Smith, J.B., jr., and Krall, K.R.: (1982), *Solar Phys.*, **80**, 233.

29 Schmieder, B., Malherbe, J.M., Mein, P., and Tandberg-Hanssen, E.: (1984), *Astron. Astrophys.*, **136**, 81.

30 Sturrock, P.A.: 1991, *Astrophys. J.*, **380**, 655.

31 Svestka, Z., and Cliver, E.W.: (1992), in *Eruptive Solar Flares*, Svestka, Z., B.V. Jackson and M.E. Machado (eds.), Springer-Verlag, New York, p. 1.

32 Menzel, D.H. and Evans, J.W.: (1953), Acad. Naz. Lincei. Convergo Volta, **11**, 119

33 van Ballegooijen, A.A., and Martens, P.C.H.: (1989), *Astrophys. J.*, **343**, 971.

34 Van Tend, W., and Kuperus, M.: (1978), *Solar Phys.*, **59**, 115.

ECLIPSE E-CORONA RECORDING AT FAR DISTANCES

I.S. KIM
Sternberg State Astronomical Institute, Moscow State University
13, Universitetsky pr., 119899 Moscow, Russia

Abstract Problems of the E-corona spectra recording in the range 2-4 $R_\odot$ are discussed. The green (Fe XIV, 5303 Å) and near infra-red (Fe XI, 7892 Å) coronal lines are concluded to be more preferable for obtaining the eclipse Fabry-Perot interferograms. The expected intensities of these lines at 3 $R_\odot$ are estimated to be 3.5-4 orders of magnitude less as compared with the intensity at 1.1 $R_\odot$. Analysis of the stray light of the Fabry-Perot eclipse instrumentation is made using former eclipse data. It's noted that successful recording the E-corona at the range 2-4 $R_\odot$ could be made by an improved eclipse instrumentation with significantly reduced level of the stray light.

1. Introduction

The most part of current eclipse projects is aimed at searching the regions of the solar wind acceleration and observational evidences for the theories of coronal heating mechanisms. Recently Grall et al.[4] indicated that the acceleration of the polar wind almost complete by 10 $R_\odot$. They emphasized the value of observations in the range 2-4 $R_\odot$. During total solar eclipses these coronal regions become easily accessible for investigations. Direct E-corona (emission line) spectra could shed some light at the problems. However, recording the coronal spectra at distances more than 1.4 $R_\odot$ is complicate till now because of weakness and steep gradient of the E-corona emissions. Advantage in light gathering power of the Fabry-Perot interferometer, or its modification as a Fabry-Perot etalon, led to a wide employment of the Fabry-Perot instrumentation (an etalon blocked by a narrow passband interference filter) for investigations of low brightness astronomical objects, particularly, the solar corona. The so-called "image-spectra" (interferograms) of the E-corona are similar to multi-slit spectra. Such eclipse interferograms are successfully obtained since 1953 [1, 2, 3, 6, 7, 15, 16, 17, 18, 19]. As a rule, astronomical or technical films were used as detectors. The above-mentioned authors reported on coronal spectra reaching 1.5 $R_\odot$. Employment of an image intensifier with a factor of 10 combined with Kodak 103aG film [15] allow to detect at 1.9 $R_\odot$ parts of interference fringes [9, 10]. The improved eclipse Fabry-Perot instrumentation should be developed to get the range 2-4 $R_\odot$. Several questions seem to be important for the problem:

- correct choice of the coronal line;
- the expected absolute intensities of coronal lines at 2-4 $R_\odot$;
- the level of the instrumental scattering.

Z. Mouradian and M. Stavinschi (eds.),
Theoretical and Observational Problems Related to Solar Eclipses, 159-163.
©1997 *Kluwer Academic Publishers.*

2. Choice of the Coronal Line

The most intensive green and red coronal lines located in the visible spectral interval (Fe XIV, 5303 Å and Fe X, 6374 Å respectively) are normally used for eclipse interferograms. Employment of these lines can be explained both by their relatively high intensity and by available detectors (films).

The correct choice of the coronal line could facilitate the task for getting the E-corona spectra at far distances. The choice of the line is usually based on comparative analysis of intensity, the sky brightness, available detectors and reliability of the interpretation of interferograms. The first three points were widely discussed before. The reliability of interpretation of Fabry-Perot interferograms depends strongly on the coronal line used because of disadvantages of the Fabry-Perot instrumentation. In particular, preliminary monochromatization (blocking) can not be done perfectly by a narrow pass band interference filters [11]. White light coronal and "cold" prominence and chromosphere emissions can penetrate through the "interference filter + etalon" combination and would be revealed in the coronal interferograms or filtergrams as traces of "false" coronal features [5]. The list of coronal lines located in blue, visual, and near infrared spectral interval was analyzed [14]. The green (Fe XIV, 5303 Å) and near infrared (Fe XI 7892 Å) coronal lines are concluded to be more preferable for reliable interpretation of coronal interferograms.

3. Expected Intensities of Coronal Lines

The green line interferograms of the solar corona of July 31, 1981 are used to estimate the absolute intensities at the range 2-4 $R_{\odot}$. The details of observations and data reduction were presented somewhere else [8]. Absolute calibration was based on the solar disk exposed during partial phases of totality through neutral density filters placed before the entrance aperture. On the contrary with former researches the secondary images of the solar disk caused by the uncoated surfaces of the Fabry-Perot plates were used as a standard. Their intensity equals to 0.03 of the solar disk one. A correction for changing the disk center altitude was made. On the contrary with employment of the main image of the solar disk the above-mentioned procedure of absolute calibration results in advantage of about 2 orders of magnitude in brightness difference between the corona and a standard. Finally, the resulted accuracy of absolute calibration is believed better 50%.

There is similarity between the distributions of the green line and white light corona brightness. The most far fringes are seen in the regions of helmets. The line-to-continuum intensity ratio seems to improve with distance. Dependences of the green line absolute intensity on distance in the range 1.1-2.0 $R_{\odot}$ are shown in Figure~1 for several white-light regions. Intensities in polar regions (plumes and coronal holes) are 1.5-2 orders of magnitude less as compared with streamers and helmet structures.

On the average the green line intensity within the range 1.1-2 $R_{\odot}$ of white-light helmet structures drops in 2 orders of magnitude. The expected average value of the green line flux at 3 $R_{\odot}$ found by extrapolation of the dependences observed equals

3 10^{-2} erg/(cm²•s•sr) being 3.5-4 orders of magnitude lower as compared with the intensity at 1.1 R$_\odot$. It means the employment of low-scattered light instrumentation.

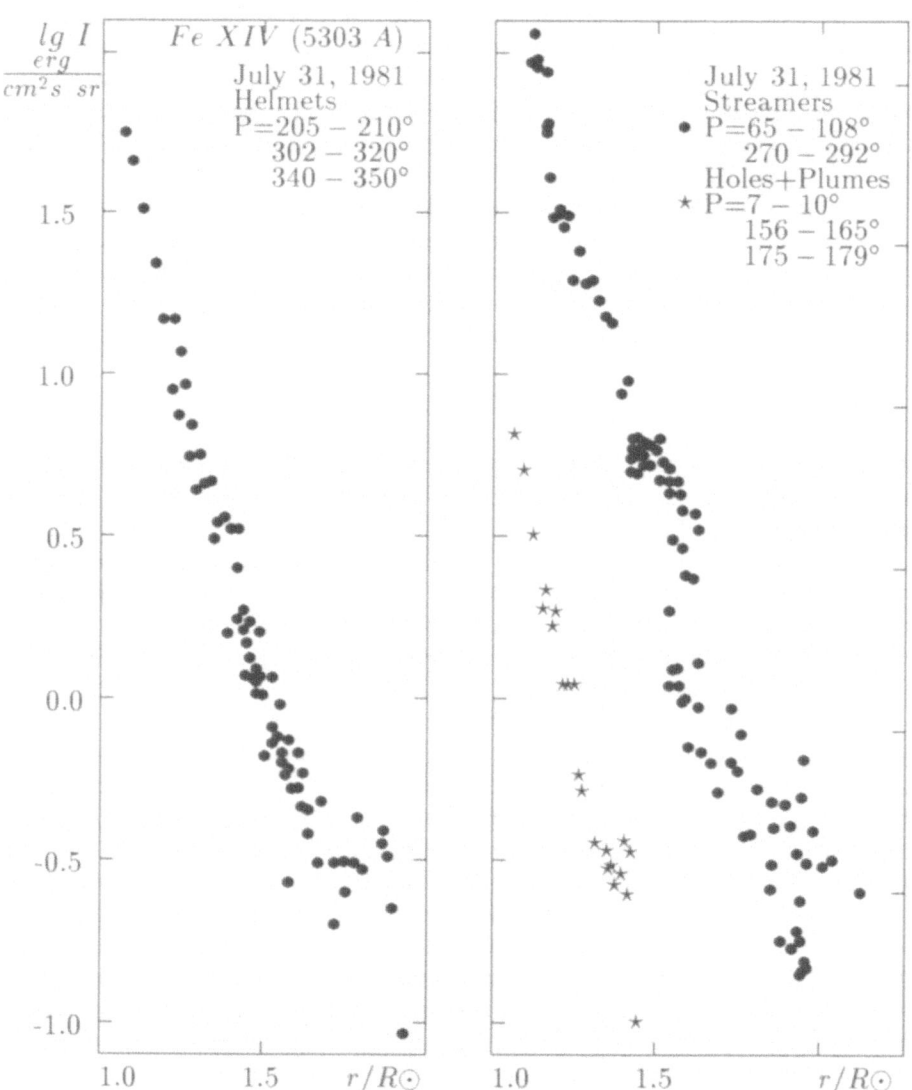

Fiugure 1. Intensity of the green line (Fe XIV, 5303Å) versus distance.

4. Stray Light of the Eclipse Fabry-Perot Instrumentation

The similar optical sketch was used for the above-cited observations. Conventional camera lenses and doublets were used as relay optics.

As a rule the over-exposed eclipse interferograms revealed parts of interference fringes imposed on the image of the Moon disk. These fringes are caused by the inner corona light scattered by the optics of the Fabry-Perot instrumentation. An over-exposed 30 s interferogram of the green line obtained by Prof. G.M. Nikolsky during the total solar eclipse of 1981 [15] is used to estimate the stray light for the conventional Fabry-Perot instrumentation. Several photometric cuts crossing the Moon disk were made. The stray light at $1.2 \, R_\odot$ equals $2 \, 10^{-2}$.

The steep E-corona gradient together with relatively high level of the stray light would prevent to record the coronal emissions at $3 \, R_\odot$. Special techniques to reduce the level of the stray light could be used [12]. Coronagraphic technique combined with a super-smooth primary optics would reduce the stray light up to 10^{-4} at least allowing to record the E-corona at $3 \, R_\odot$.

5. Conclusions

The green (Fe XIV, 5303 Å) and near infra-red (Fe XI 7892Å) coronal lines seem to be more preferable to get direct spectra in the range 2-4 $R_\odot$ by a Fabry-Perot instrumentation.

A low-scattered light primary optics combined with the Lyot stops are recommended to be used to reduce the stray light. A radial filter to reduce the intensive inner corona scattered light can be placed in the secondary focal plane.

Fabry-Perot etalon mirrors and interference filters are recommended to be placed in the plane of Lyot's stop.

The version of such instrumentation have been tested in Brazil during the total solar eclipse of 03 November 1994. Two double pass coronagraphs with achromatic lenses as primary optics were attached by the Fabry-Perot etalons, interference filters and image intensifiers. The interferogram obtained by one of them (observations in Chapeco town) reveals the parts of fringes at 2.3 $R_\odot$. The interferogram is being analyzed now to take into account the stray light which is still relatively high. The forthcoming total solar eclipse of 1999 gives an opportunity to try to record the coronal spectra at far distances. The two above-mentioned instruments could be used. Replacement of the primary optics as well as detectors would improve the instruments. Improvement of the instrumentation in the frame of international collaboration, in particular, in respect to the detector assemblies (CCD) would allow observations at two observational sites placed apart from each other to get information about coronal dynamics.

Acknowledgements

The author thanks the organizers of the NATO ARW supporting the attendance the workshop. This work has been supported by the grants 94-02-2876 and 96-02-17753 of Russian Bound for Fundamental Researches and partly by the grant 93-8-229 of the Russian State Program "Astronomy".

References

[1] Delone, A.B., Makarova, E.A. (1969) Interferometric investigations of the red and green coronal lines during the total solar eclipse of May 30, 1965, *Solar Phys.*, **9**, 116–130

[2] Delone, A.B., Makarova, E.A., Yakunina, G.V. (1988) Evidence for moving features in the corona from emission line profiles observed during eclipses, *J. Astrophys. Astr.*, **9**,41–47

[3] Desai, J.N., Chandrasekhar, T. (1983) Temperature distribution in the corona from 5303 A line width observations. Eclipse of 1980 February16 - Tentative evidence for a temperature maximum, *J. Astrophys. Astr.***4**, 65-74

[4] Grall, R.R., Coles, W.A., Klinglesmith, M.T., Breen, A.R., Williams, P.J.S., Markkanen, J., and Esser, R. (1996) Rapid acceleration of the polar solar wind, *Nature*, **379**, Letters, 429-432

[5] Guetman, F.I., Kim, I.S., Bucher, A., Druzhinin, C.A., Noens, J.-C.,Salakhutdinov, R.T., Semenikin, A.A., and Skomorovsky, V.I. (1994) On filter observations of coronal structures in the light of Fe X 6374 Å emission line, *Proceedings of IAU Colloq. 144 "Solar Coronal Structures"*, V. Rusin, P. Heinzel, and J.-C. Vial (eds.), VEDA Publishing company, Bratislava - printed in Slovakia, 575-578

[6] Hirshberg, J.G., Wourtes, A., Hazelton, L. (1971) Interferometric studies of spectral lines in the solar corona, *Solar Phys.*, **21**, .448-451

[7] Jarret, A.H., Kluber, H. (1955) Interferometric measurements of the green coronal line during the total solar eclipse of 1954, June 30, *Monthly Notices Roy. Astron. Soc.*, **115**, 343-362

[8] Kim, I.S. (1992) On the Fabry-Perot investigations of the solar corona: eclipse observations of the green line intensities and half widths, *ESA SP-348*, 149-152

[9] Kim, I.S. (1994) Eclipse observations of large-scale coronal dynamics: coronal rotation, *Proceedings of the seventh European Meeting on Solar Physics "Advances in Solar Physics"* Pubblicazioni dell 'Osservatorio Astrofisico di Catania, Belvedere, G., Rodono, M., Schmieder, B., and Simnett, G. (eds.), Arti Grafishe, Signorello, Catania ·, 141-146

[10] Kim, I.S. (1994) On the Fabry-Perot investigations of the solar corona: Eclipse observations of large-scale dynamics, *Adv.Space Res.***14**, (4)45-(4)48

[11] Kim, I.S., Alexeeva, I.V., and Smartt, R.N. (1994) Comments on Doppler shifts deduced by the Fabry-Perot technique, *ESA SP-373*, December 1994, 71-73

[12] Kim, I.S., Bougaenko, O.I., Brouevitch, V.V., and Evseyev, O.A.(1995) Problems of reflecting coronagraphs, *Izvestiya Rossiskoi Akademii Nauk, ser. fizicheskaya*, **43**, 153-160

[13] Kim, I.S. and Nikolsky G.M. (1975) Investigation of emission lines ofthe solar corona of 10 July, 1972 using the Fabry-Perot etalon, *Solar Phys.*, **43**, 351-358

[14] Kim, I.S., and Smartt, R.N (1994b) On the eclipse E-corona observations by the Fabry-Perot technique, *Proceedings of IAU Colloq. 144 "Solar Coronal Structures"*, V. Rusin, P. Heinzel, and J.-C. Vial (eds.), VEDA Publishing company, Bratislava - printed in Slovakia , 549-553

[15] Koutchmy, S. and Nikolsky, G.M, (1982) Soviet-french observations of the total solar eclipse, *Zemlya i Vselennaya*, 65-67

[16] Liebenberg, D.H., Bessey, R.J., Watson, B. (1975) Observed coronal temperatures at 1.37 Ro in the region of a helmet structure, *Solar Phys.*,**40**, 387-396

[17] Marshall, P.M. and Henderson G. (1973) An interferometric investigation of emission lines of the solar corona, *Solar Phys.*, **33**, 153-168

[18] Raju, K.P., Desai, J.N., Chandrasekhar, T., Ashok,N.M. (1993) Line-of-sight velocities observed in the inner solar corona during the total solar eclipses of 1980 and 1983, *Monthly Notices Roy. Astr. Soc.*,**263** ,789-797

[19] Smartt, R.N., Zirker, J.B., and Mauter, H.A. (1982) Interferometric eclipse observations of the Fe XIV inner corona, *Proc. Indian Natn. Sci. Acad.*, **8** A, Supplement No~3,102-108

SIMULATION METHOD TO COMPUTE ACCURATELY THE EINSTEIN EFFECT DURING A TOTAL SOLAR ECLIPSE. NEW IDEAS.

H. DEBEHOGNE
Observatoire Royal de Belgique,
Av. Circulaire 3, B-1180 Brussels,   Belgium

Abstract.
Simulation, fictitious and test stars, matrix language and multiplication, distorsion formula, α and δ separated in accuraccy studies, true error (e) computed from residual (R), error effect, are presented.

1. Introduction.

The discordences between computed Einstein Effect's [8] leave to new methods. SIMULATION method has been applied at the March 7, 1970 eclipse, observed at Nantucket, USA, by D. Hoffleit [6] from the Yale University.

2. Methods of Simulation.

The theory furnishs as deflection

$$\Delta R = 1."745/R,$$

R being the distance to the solar center expressed in solar radius. Two plates, at eclipse and without the sun (comparison), are taken.

    1. Both plates are reduced. Various EE (Einstein Effect) are applied at eclipse.The good EE is the one furnishing the same residuals on both plates.

    2. Fictitious Stars (FS) are determined on the comparison. At eclipse, various EE are applied to the basis fictitious stars. The right value is the one giving the the accurate coordinates of the catalogued stars.

    3. Various EE are applied to the real stars on the comparison. The right EE is the one deducing the same sky fictitious network from both plates.

3. Formulas of Reduction. Matrix Language

The bijection (a_{ij}, b_{ij}) between plate (x,y: measurements) and sky (X,Y: celestial rectilinear or standard coordinates) is determined by two INDEPENDENT POLYNOMIAL EQUATIONS [3]:

Z. Mouradian and M. Stavinschi (eds.),
Theoretical and Observational Problems Related to Solar Eclipses, 165-168.
©1997 *Kluwer Academic Publishers.*

$$X = \sum_{i+j=k=0}^{n} a_{ij}\, x^i\, y^j, \qquad Y = \sum_{i+j=k=0}^{n'} b_{ij}\, x^i\, y^j \qquad (1)$$

i, j integers, i decreasing, j increasing in $i+j = k$; $N = (n+1)(n+2)/2$ unknowns a_{ij}, $N' = (n'+1)(n'+2)/2$ unknowns b_{ij}.

Owing to the bijection between (α,δ) and (X, Y) [3] the M ($\gg$N) basis stars for X, M' ($\gg$N') basis stars for Y, give two independent rectangular systems of M and M' equations containing N and N' unknowns with M and M' independent terms in the vectors (X) and (Y).

The two systems are written

$$(S\ x^i\ y^j).v(a_{ij}) = (X) \qquad (2)$$

$$(S'\ x^i\ y^j).v(b_{ij}) = (Y) \qquad (3)$$

$(S\ x^i\ y^j)$ and $(S'\ x^i\ y^j)$ being two rectangular matrixes of M, M' lines and N, N' columns (M $\gg$ N), (M' $\gg$ N').

Each rectangular system becomes square system by multiplying both sides, respectively by $T(S\ x^i\ y^j)$ and $T(S'\ x^i\ y^j)$ the transposed matrixes of (Sx^iy^j) and $(S'x^iy^j)$ [7].

Both square systems are solved by one of the well known methods.

4. General Representation of Various Bijections

See Debehogne[5].

5. Better Proposed Formulas

1. The bijection of degree 3 (used for March 7, 1970)

$$X = a_{00} + a_{10}\, x + a_{01}\, y + a_{20}\, x^2 + a_{11} x\, y + a_{02}\, y^2 + \\ + a_{30}\, x^3 + a_{21}\, x^2\, y + a_{12}\, xy^2 + a_{03}y^3$$

$$\qquad (4)$$

$$Y = b_{00} + b_{10}\, x + b_{01}\, y + b_{20}\, x^2 + b_{11} x\, y + b_{02}\, y^2 + \\ + b_{30}\, x^3 + b_{21}\, x^2\, y + b_{12}\, xy^2 + b_{03}\, y^3$$

0 term omitted in X and Y, is simply written

$$(3,0,0). \qquad (5)$$

2. Degree 3, incomplete, the best for ESO-Schmidt and for EE,

$$X = a_{00} + a_{10} x + a_{01} y + a_{20} x^2 + a_{11} x y +$$
$$+ a_{30} x^3 + a_{12} xy^2$$

(6)

$$Y = b_{00} + b_{10} x + b_{01} y + b_{11} x y + b_{02} y^2 +$$
$$+ b_{21} x^2 y + b_{03} y^3$$

3 terms (6,8,10) omitted in X, 3 terms (4,7,9) omitted in Y, is

$$(3,3,3;6,8,10;4,7,9).$$

(7)

This is the "1rst degree distorsion" [2,3,5].

6. Test Stars

See Debehogne [2,3,5].

7. Example of Bad and Right Accuracy Studies

Star	R_α	R_δ	$\sqrt{R_\alpha^2 + R_\delta^2}$
1	0.0	3.0	3.00
2	3.0	0.0	3.00
3	2.5	2.5	3.53

Bad Elimination : star, 3, for α and δ by considerationof the greatest

$$\sqrt{R_\alpha^2 + R_\delta^2}$$

Right Elimination: star 2 is put away from the system for X (but not out of the system for Y) and star 1 is put away from the system for Y (but not out of the system for X), it is to say, SEPARATELY, in α and δ.

But, "right" is not sure because the star with the greatest residual is not always erroneous. Only the construction of the equiresiduals lines on the plate gives the bad star(s) and the error. The ERROR SIMULATION permits the same [7].

8. Accuracy : Error Replaces Residual

We know the relation

$$|e| = |R| + |ee|$$

(8)

e = error, R = residual, ee = error effect.

The difference between the coordinate after an error simulation (S) and its previous value is the Simulation's effect (Se). We suppose and then we can deduce

$$\frac{e}{ee} = \frac{S}{Se}$$

$$\frac{e}{e - ee} = \frac{S}{S - Se}$$

or, by (8),

$$e = R \frac{S}{S - Se} \tag{9}$$

If the relation between error and error-effect is not linear, an appropriate expression can be deduced as multiplier of R. In the EE, a possible error is to take it equal to R when it is e.

9. Error Effects

See Debehogne [1,5]. Here too, the EE could be taken as e when it is ee.

10. Conclusions

The observations of the eclipses in future must be done. The measurements must be published for future new methods. Simulation and fictitious stars avoid the contingencies of real basis.

Acknowlegments

The team of Miss D. Hoffleit, University of Yale, Massachusetts, USA, has taken both plates (eclipse and comparison) at the Observatory of Nantucket for the Total Solar Eclipse of March 7, 1970, and sent them graciously in Belgium to the author.

References

1. Debehogne, H. (1969), *Bull. Ac. Roy. Belg. Cl. S.*, tome **LV**, 5, 415.
2. Debehogne, H. (1970), *A&A* **8**, 189.
3. Debehogne, H. (1970), *Cospar Transactions, Prague* 1968, 7, 107.
4. Debehogne, H. (1972), *A&AS*, **5**, 185.
5. Debehogne, H. (1974), *A&AS*, **16**, 353.
6. Debehogne, H. (1977,) *Bull. Ac. Roy. Belg. Cl. S.*, tome LXIII, 5, 165.
7. Debehogne, H. (1986,) *Acta Astronomica*, *36*, 301.
8. Mérat, P., Pecker, J.C., Vigier, J.P., Yourgrau, W., (1974), *A&A*, **32**, 471.

EMISSION LINE SPECTRUM OF THE SOLAR CORONA

Z. MOURADIAN
Observatoire de Paris-Meudon
92195 Meudon Cedex, France

Abstract: This paper reviews the recent eclipse observations of emission lines in the visible and near infrared coronal spectrum (300 to 3000 nm). The ion identifications and intensity measurements are discussed, and a number of guidelines for future observations are suggested.

1. Introduction

The identification of coronal lines becames possible when it has been understood that the temperature of the corona is much higher than that of the photosphere, and that consequently the atoms must be ionized. In 1929, Grotrian observed that the scattered photospheric spectrum in the range of CaII H and K lines are very smooth in the corona. Lyot, in 1932, measured the line width of the green line and found 0.9 Å. A littele later, Fe VI and Fe VII lines were detected in stellar spectra. All these observational facts indicate that the temperature of the corona is very high and, consequently, that the atoms are highly ionised. By 1942, 19 coronal lines had already been identified [6]. Before the Second World War, the problem in coronal physics was, How are the coronal spectrum lines formed? Since the war, the question has been, Why is the temperature so high? This question remains unresolved.

The spectrum of the Solar corona reflects a great variety of temperatures and densities, which are due to the uneven deposit of energy and the transport of the plasma, phenomena that are guided by magnetic fields. The observed spectrum therefore expresses the magnetic field geometry, the density, and the energy deposit, in the loops and arches that make up the corona.

In the present talk I shall try to approach the question of the emission line spectrum observed during the Solar Total Eclipses.

2. List of Observed Lines

Tabel I shows coronal lines in the visible and near infrared domain from 300 to 3000 nm. In the period after 1950, special care was taken to obtain slit spectrograms, instead of slitless spectrograms, in order to improve the wavelength measure of lines, which is basic for line identification. The results of following eclipse observations were considered:

Z. Mouradian and M. Stavinschi (eds.),
Theoretical and Observational Problems Related to Solar Eclipses, 169-176.
© 1997 *Kluwer Academic Publishers.*

25 February, 1952. B. Lyot built two spectrographs (one in UV and one in visible), with circular slits surrounding the solar limb. The spectrum observed was 310-690 nm [2]. This was published in the form of an atlas [3] and analysed in two articles by Divan and C. Pecker [8] and by Aly, Evans and Orrall [4]. The lines measured by Divan and C. Pecker are listed in Table I, with a complement of elements from Aly [2].

15 February, 1961. From an aircraft, Kurt [14] observed, , the spectrum from 780 to 1200 nm, with a spectrograph combined with an image converter.

15 February, 1961. Wlerick and Fehrenbach [27] observed the near infrared coronal spectrum (650-900 nm) with the $\varnothing$193 cm telescope of Haute Provence Observatory.

15 February 1961. Servajean took spectra in the 390-650 nm range, using Lyot's visible spectrograph. This spectrum was studied by Kernoa, Michard and Servajean [13].

30 May, 1965. Dunn [10] built two identical telescope-spectrographs with circular slits, in order to observe the whole coronal spectrum between 300 and 900 nm. One instrument was used on Bellingshausen Island by Jefferies, Orrall and Zirker [11]. The other was carred on a NASA aircraft, and the results were studied by Curtis, Dunn and Orrall [7] and Magnant-Crifo [16]. Later, Rohrlich and C. Pecker[24] and Svenson et al. [26], revised the ion identification.

30 May,1965. Magnus and Stockhausen [17] used a Michelson interfero-meter, aboard an airplane. They scaned the spectrum from 1 to 3.5 μ.

12 November, 1966. Münch, Neugebauer and McCammon [21] observed two spectral ranges around the Si X and Mg VIII infrared lines, using a high altitude aircraft flight.

12 November, 1966. Bayard and Kissell [5] observed the infrared Fe XIII lines with a spectrograph, aboard an aircraft.

7 March, 1970. Olson, Anderson and Steward [22] obtained the infrared spectrum between 1000 and 3000 nm, aboard an aircraft at 12 km altitude. Kastner [12] revised the suggested identifications.

7 March, 1970. Mouradian [18] obtained two spectra at the top of the chromosphere. The slit was aligned 4000 km above the photosphere and the spectrum obseved ranged from 360 to 670 nm. The spectra show chromospheric and coronal lines and coronal continuum (Figure 1).

The Table I gives the wavelength, the ion identification proposed by the authors, the transition and the ionisation potential, the ion class and in reference column, the authors. The last column notes any revision, in the identification.

Table I contain 91 spectral lines of which only 31 were observed twice or more, the level that implyes a certain security that the wavelength measures are accurate.

TABLE I Coronal lines observed during recent Solar eclipses (300-3000 nm)

λ nm	Ion	Transition	I. P. eV	Ion Class	References	Remarks
302.13	FeXII	$^2P_{1/2}$-$^2D_{3/2}$	290	I	[11]	[26]:NiXII?
307.20				I	[11]	[26]:FeXII
312.40				I or II	[11]	[26]:FeIX?
316.70	CrXI	$^1D_2$-$^3P_1$	249	II	[11]	[24]; [26]
330.28	CrIX	$^1D_2$-$^3P_2$	185	I or II	[11];[16]	[26]:NiXII
332.67					[2]	
332.75	CaXII	$^2P_{1/2}$-$^2P_{3/2}$	592	III	[11]; [16]	[26]
333.85				I?	[11]; [16]	
335.51				I?	[11]; [16]	[26]:FeIX
338.8					[8]	
338.85	FeXIII	$^1D_2$-$^3P_2$	330	II	[11]; [16]	[26]
345.4					[8]	
345.42				I	[11]; [16]	
347.1					[8]	
347.16				I	[11]; [16]	[26]:FeIX
348.85				I or II	[11]; [16]	
350.25				III	[11]; [16]	[26]:CXIII?
353.36				I	[11]; [16]	
353.5					[8]	
357.6					[8]	
357.71				I or II	[11];[16]	
360.1					[8]	
360.11	NiXVI	$^2P_{3/2}$-$^2P_{1/2}$	455	III	[11]; [16]	[26]
364.27	NiXIII	$^1D_2$-$^3P_1$	350	I or II	[11]; [16]	[26]:FeIX
364.29					[8]	
368.50					[2]	
368.55	MnXII	$^1D_2$-$^3P_2$	288	I or II	[11]; [16]	[24]:MnXII; [26]:MnXIII
380.08					[2]	[24]:CoXII
380.08	NiXI?	$^3F_2$-$^1D_2$	225	I	[16]	[26]:FeIX
380.12				I	[11]	[26]:FeIX
380.2	FeX			I	[18]	
386.93	(NeIII)			I	[18]	[1]
387.09				I?	[18]	
398.68	FeXI	$^1D_2$-$^3P_1$	262	I	[11]; [16]	[26]
398.7	FeXI		261	Ib	[13]	
398.7					[8]	
399.68	CrXI	$^1D_2$-$^3P_2$	249	I	[11]; [16]	[24]; [26]
399.71					[8]	
399.8	CrXI			I b	[13]	
399.942				I?	[18]	
403.187	NiVIII				[18]	
406.74					[2]	[1]: FIV
408.65					[8]	
408.71	CaXIII	$^3P_1$-$^3P_2$	655	III	[11]; [16]	[26]
411.1				III b	[13]	
422.27					[8]	
423.105	NiXII			II	[18]	
423.12	NiXII	$^2P_{1/2}$ - $^2P_{3/2}$	321	II	[11]; [16]	[26]
423.2	NiXII				[13]	
423.2					[8]	

423.204	?				[18]	
427.3 ?					[13]	
431.18				II	[11]; [16]	
431.2					[2]	
432.2				I b	[13]	
435.1	Co XV	$^2P_{3/2}$ -$^2P_{1/2}$		III b	[13]	[24]; [26]
435.87					[8]	
435.9	Fe X			I b	[13]	
435.857	Fe IX			I	[18]	
435.94				I	[11]; [16]	[26]:FeIX
436.3 ?					[8]	
441.2					[2]	
441.2	Ar XIV		682	III b	[13]	
441.2	Ar XIV	$^2P_{3/2}$ -$^2P_{1/2}$	687	IV	[11]; [16]	[24]; [26]
441.7 ?					[8]	
444.9	?				[18]	
448.4					[8]	
456.6	?				[18]	
456.61					[2]	
456.62				I	[11]; [16]	
456.7	CrIX			I	[13]	[1]:$^1 D_2$ -$^3P_1$
457.65					[2]	
458.53				II	[11]; [16]	[26]:FeIX
458.57					[2]	
458.6				I b?	[13]	
511.48					[2]	
511.58	NiXIII	$^3P_1$ -$^3P_2$	350	II	[11]; [16]	[26]
511.6	NiXIII		350	II?	[13]	
530.275	FeXIV			II	[18]	
530.3	FeXIV				[8]	
530.3	FeXIV		355	II	[13]	
530.34	FeXIV	$^2P_{3/2}$ - $^2P_{1/2}$	355	II	[11]; [16]	[26]
535.962				I?	[18]	
537.1	FeVI?	$^2G_{7/2}$ -$^4F_{9/2}$	110		[18]	
544.4	CaXV	$^3P_2$ -$^3P_1$	820	IV	[11]; [16]	[24]; [26]
544.5	CaXV		814	IV	[13]	
544.6					[2]	
553.34				I	[11]; [16]	[26]:ArX
553.9	ArX		421	II?	[13]	
553.91	ArX	$^2P_{1/2}$ - $^2P_{3/2}$	423	I	[11]; [16]	
553.95					[2]	
569.36	CaXV	$^3P_1$ - $^3P_0$	820	IV	[11]; [16]	[24]; [26]
569.37					[2]	
569.4	CaXV		814	IV	[13]	
627.206					[27]	
630.460					[27]	
637.4	FeX				[8]	
637.4	FeX		233	I	[13]	
637.4	FeX				[18]	
637.456	FeX				[27]	
637.46	FeX	$^2P_{1/2}$ -$^2P_{3/2}$	235		[11]; [16]	[26]
643.60	(MnXIII)				[27]	
653.596	MnXIII				[27]	
653.63	MnXIII	$^2P_{3/2}$ -$^2P_{1/2}$	315	I or II	[11]; [16]	[24]; [26]

670.147	NiXV				[27]	
670.17	NiXV	$^3P_1 - ^3P_0$	430	III	[11]; [16]	[26]
691.7	AXI?				[27]	[26]:ArXI
705.959	FeXV				[27]	
705.86	FeXV	$^3P_2 - ^3P_1$	390	III	[11]; [16]	[26]
709.2.					[27]	
714.390					[27]	
716.075					[27]	
721.276					[27]	
761.10	SXII	$^2P_{3/2} - ^2P_{1/2}$	506	II	[11]; [16]	[26]
7852..0				(III)	[27]	
789.2	FeXI				[14]	
789.18	FeXI	$^3P_1 - ^3P_2$	262	I	[11]; [16]	[26]
789.189	FeXI				[27]	
795.6					[14]	
802.41	NiXV	$^3P_2 - ^3P_1$	430	II or III	[11]; [16]	[26]
807.7					[27]	
815.38	CrXII	$^2P_{3/2} - ^2P_{1/2}$	272	I	[11]; [16]	[26]
842.508					[27]	
842.734					[27]	
842.910					[27]	
847.566					[27]	[24]:ArXIII
847.708					[27]	
849.309					[27]	
1074.68	FeXIII	$^3P_1 - ^3P_0$	330	II	[11]	[26]
1074.680	FeXIII				[5]	
1074.7	FeXIII				[14]	
1079.79	FeXIII	$^3P_2 - ^3P_1$	330	II	[11]	[26]
1079.795	FeXIII				[5]	
1079.8	FeXIII				[14]	
1097.6					[8]	[12]:FeXIV
1130.4					[14]	
1135.5					[14]	
1138.6					[14]	
1158.5					[14]	
1266.1	S IX	$^3P_1 - ^3P_{1/2}$			[22]	[26]
1430.5	SiX				[21]	
1430.8	SiX	$^2P_{3/2} - ^2P_{1/2}$			[22]	
1523.2	CrXI	$^3P_2 - ^3P_1$			[22]	
1528.1	?				[22]	
1721.5	?				[22]	[12]:TiVI
1855.6	CrXI	$^3P_1 - ^3P_0$			[22]	[12]:HeI
1921.6	SiXI	$^3P_2 - ^3P_1$			[22]	[12]:SXI
2747.3	AlX	$^3P_2 - ^3P_1$			[22]	
3019.3	MgVIII	$^2P_{3/2} - ^2P_{1/2}$			[22]	[26]:3027.5
3027.5	MgVIII				[21]	

3. Ion Classes.

The first step in the identification process, is to classify the spectral lines, which may indicate the ion's domain of ionisation potential. The intensity variation along thespectrograph slit differs to a greater or to a lesser degree from one line to another, depending on the variable ion densities along the line of sight. The ions can be arranged in four groups, denoted I to IV, by order of increasing ionisation potential. Aly, Evans and Orrall [4] illustrate well the intensity fluctuations around the Solar limb for the quiet Sun and for the corona above an active region (coronal condensation). In the latter case, all four classes are present.

Ion classes can be interpreted roughly in terms of temperature or ionisation potential. The ion classification was first proposed by Lyot [15] and it found its definitive form with Shajn [25] and with Dollfus [9]. Below, for each class, we give a representative line and limits of ionisation potential:

I 637.4 nm	II 503.3 nm	III 670.2 nm	IV 569.4 nm
150-300 eV	300-400 eV	400-700 eV	700-900 eV

Class I lines show relatively cool coronal regions close to the transition to the corona from the chromosphere or prominences. Class II lines are related to active regions and to moderate solar activity. The hotter parts of active regions emit in class III and IV lines.

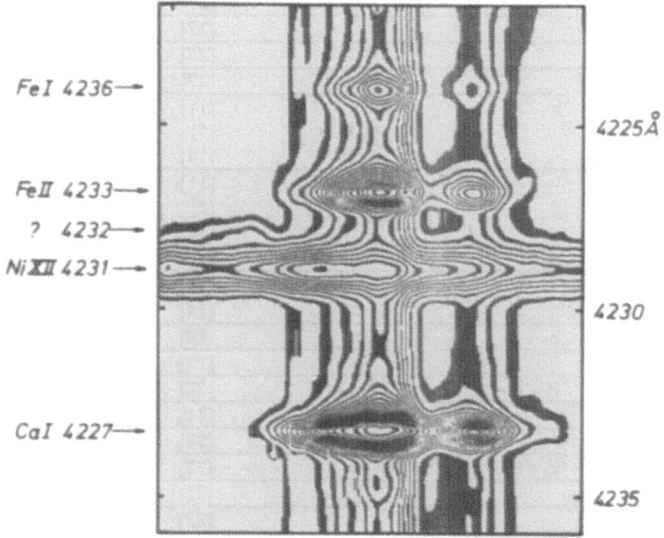

Figure. 1 Two spectra showing chromospheric lines, coronal continuum, and coronal lines. The ? denotes unidentified coronal lines observed at the total Solar eclipse of 7 March 1970 [18].

4. Identification of Coronal Lines

Coronal lines are formed either by magnetic dipole or electric quadrupole transitions. In Table I the magnetic dipoles are double (22 lines) and are related to high ionisation potentials (IP >235 eV), whereas the elecrtoic quadrupoles (10 lines) are related to low ionisation potentials (IP <350 eV).

Generally, coronal line emissions result from transitions to the ground level. Essentially, the magnetic dipole results from highly ionised atoms for which the nucleus produces a splitting of ground states, whose transition ($\Delta L = 0$ and $\Delta J = \pm 1$) are in visible or near IR spectrum. A typical example is the green line, (FeXIV):

λ 530.3 nm $3p$ $^2P_{3/2}$-$^2P_{1/2}$

On other hand, the Fe XIII line shows both kinds of forbidden transitions:

λ 338.85 nm $3p^2$ $^1D_2$-$^2P_2$
λ1074.68 nm $3p^2$ $^3P_2$-$^3P_0$
λ1079.79 nm $3p^2$ $^3P_2$-$^3P_1$

We remark that coronal lines exist in all isoelectronic sequences except those of H and He, for which the transitions are in the far infrared [23].

5. Coronal Line Intensities

The measures of intensity of coronal lines have been less successful than wavelength measurements. This problem is because of the difficulty of knowing exactly where the slit is positioned with, respect to the hidden Solar limb, and because of the errors in absolute photometric standardisation. The hot active regions are clearly visible in classes II-IV, whereas the quiet Sun is well visible in class I lines.

Since the height variation of spectral lines, may indicate the excitation mechanism (collision or radiation), efforts were made in the article of Jefferies et al. [11], to provide absolute spectral line measurements as function of height. The article of Magnan-Crifo [16] also containts measurements as a function of altitude.

6. Conclusions and Recommendations

We would lick to conclude with a few recommendations for future eclipse observations:

i) The first is a reaction to the fact that the observations of the coronal emission spectrum stopped once most of the prominent lines were correctly observed. But it should be noted that many lines have been observed only once, especially the weaker ones. An effort should to improuve this situaion.

ii) Line identification analysis has also stopped, with only 42 ont of 91 lines determined, even if we grant that all reported identifications are actually correct. We would like to point out the need for a general approach to the identification problems, which is a very complex, but useful program. Table I gives many lines that have not been identified. There is still much to de done in detection and identification of ions with low IP.

iii) As concerns the interpretation of coronal observations, more attention should be paid to the chromospheric structures underlying the observed corona. To help in this kind of analysis, Mouradian et al.[19, 20] gave special synoptic charts for certain past eclipse observations (1990, 1991 and 1994). Chromospheric morphological analysis may explain some of the spectral line intensity variations occurring along the solar limb.

iv) Care should be taken in equivalent-width measurements, because of the influence of the instrumental polarisation on the intensity of the K corona. Certain eclipse instruments act as partial polarizers.

v) It is obvious that the height variation of spectral-line intensity in absolute units would be a very useful observation.

References:

1. Allen,C.W. (1973,) *Astrophysical Quantities*, third ed., The Athlone Press, London
2. Aly,M.K. (1955) Preliminary note on measures of Coronal emission lines observed at the total Solar eclipse, February 25,1952, by B. Lyot and M.K.Aly *Ap.J.* **122**, 438.
3. Aly,M.K., Evans,J.W., Orrall,F.Q. (1962,) *An Spectrographic Atlas of the Solar Corona*, Helwan Observatory and Sacramento Peak Observatory
4. Aly,M.K., Evans,J.W., Orrall,F.Q. (1962) A photometric study of the continum and seventeen emission lines in the inner Solar Corona *Ap.J.* **136**, 956
5. Bayard,P.L., Kissell,K.E. (1971) Observations of the infrared Fe XIII lines in the Solar Corona of 12 November, 1966 *Solar Phys.* **21**, 351
6. Billings, D. E. (1966) *A Guide to the Solar Corona*, Academic Press, New York
7. Curtis, G.Wm., Dunn, R.B., Orrall, F.Q. (1965) *Proceedings 1965 Solar Eclipse Symposium*, NASA Ames Res.Center, 137
8. Divan,L., Pecker,C. (1960) Etude des spectres pris par Bernard Lyot à L'éclipse de Khartoum du 28 février 1952, *Ann.d'Astroph.* **23**, 541
9. Dollfus,A. (1957) Etude des spectres de la Couronne solaire obtenus à Khartoum pendant l'éclipse totale de Soleil du 25 février 1952, *C.R.Acad.Sc.Paris* **245**, 2011
10. Dunn,R.B. (1966) A gyro-stabilized air-borne eclipse spectrograph, *Sacramento Peak Observatory Contrb.*No .97
11. Jefferies,J.T., Orrall,F.Q., Zirker,J.B. (1971) The spectrum of the inner Corona observed during the total Solar eclipse of 30 May 1965, *Solar Phys.* **16**, 103
12. Kastner, S.O.:1993 Re-analysis of an observed Solar near-infrared eclipse spectrum, *Solar Phys.* **143**, 197
13. Kernoa,A., Michard,R., Servajean,R. (1965) Contribution à l'étude de la Couronne lors de l'éclipse du 15 février 1961. IV. Photométrie des raies d'émission de la Curonne *Ann d'Astroph.* **28**, 716
14. Kurt, V.G.(1962) An investigation of the spectrum of the solar corona in the wavelength region 7800-12000A during the total solar eclipse of February 15, 1961, *Soviet Astronomy-AJ* **6**, 349
15. Lyot, B. (1939) A study of the Solar Corona and prominences without eclipse, *M.N.R.A.S.* **99**, 580
16. Magnant-Crifo,F. (1973) The visible spectrum of the lower corona during the total eclipes of May 30; 1965;*Solar Phys.* **31**, 91
17. Magnus, J., Stockhausen, R. (1966), New infrared Solar eclipse observations, *Proceedings 1965 Solar Eclipse Symposium*, NASA Ames Res.Center,137, NASA, Ames Res. Cent., 329
18. Mouradian,Z. (1972) Raies nouvelles observées lors de l'éclipse du 7 mars 1970, *Solar Phys.* **24**, 368
19. Mouradian,Z., Buchholtz,G., Zlicaric,G. (1994) Synoptic charts of the Solar eclipse of 1990 and 1991 in *Solar Coronal Structres*, V.Rusin, P.Heinzel, J.C.Vial, (eds), VEDA Pub. Co., 517
20. Mouradian,Z., Prevot,G., Zlicaric,G., Garcia,A.G. (1995) Chromospheric synoptic charts for the Solar eclipse of november 3, 1994, *Solar Phys.* **158**, 379
21. Münch,G., Neugebauer, G., McCammon, D.(1967) Infrared coronal lines. II Observation of [Si X] λ 1.43μ and [Mg VIII] λ3.03μ *Ap.J.* **149**; 681
22. Olsen,K.H., Anderson,C.R., Stewart,J.N.(1971) Some newly discovered coronal emission lines from high altitude infrared observations of the 7 March, 1970, solar eclipse; *Solar Phys.* **21**, 360
23. Pryce,M.H.L. (1964) The origin of Coronal emission lines, *Ap J.* **140**, 1192
24. Rohrlich,F., Pecker,C. (1963) Highly ionized atoms: the configuration : $s^2 p^n$ (n=1,2,4,5), *Ap.J.* **138**, 1247
25. Shajn,G.A. (1948) On the intensity emission lines of the coronal spectrum, *Izv. Krymsk. Ap. Obs.* **3**, 64
26. Svansson,L.A., Ekberg,J.O., Edlén,B.:(1974) The identification of Fe IX and Ni XI in the SolarCorona, *Solar Phys.* **34**, 173
27. Wlerick,G., Fehrenbach,C. (1963) Premier résultats cocernant les spectres infrarouges de la Couronne obtenus pendant l'éclipse du 15 février 1961,in J.W.Evans (ed.), *The Active Regions Solar Corona*, IAU Symph. No. 16, Academic Press, p. 199

A POSSIBILITY FOR STUDYING SUBTELESCOPIC STRUCTURES IN THE SOLAR CORONA

I.P. KRYACHKO, L.N. KUROCHKA AND
V.V. TEL'NUK-ADAMCHUK
Astronomical Observatory of the Kyiv, Taras Shevchenko University
3 Observatorna Str., UA-254053 Kyiv, Ukraine

Abstract. It is shown that information on the existence of subtelescopic (unresolved) inhomogeneities having increased density in the coronal structures may be obtained by comparing two filtergrams one either side of the Balmer jump, one at $\lambda < 365$ nm and the other at, for example $\lambda = 580$ nm or $\lambda = 477$ nm .

1. Introduction

Solar eclipse observations can be used not only for the study of corona structure but also for determining the change of the coronal electron density with height. This height is obtained from measuring the observed fall-off of the coronal emission intensity with distance from the solar limb. The average model of Baumbach [1] for the solar corona obtained with this approach is well known [2, 3].

It should be bor§§e in mind that the electron density (n_e) distribution in the corona is assumed to decrease monotonically with distance from the Sun, while there is the coronal emission by the Thompson scattering of solar radiation by electrons. If the solar corona is structurally inhomogeneous on the subtelescopic level, and consists, for example, of low-density inhomogeneities (condensations or clots), then an agreement exists with observations [4]. The intensities of the coronal lines could be explained by this. This equally applies to the fall off of intensity with a distance in prominence emission and limb flares. The question of what kind of emission formation i.e. about the model (a model with monotonic decrease in the electron density or one having a decrease in the number of subtelescopic inhomogeneities outward) can be solved with new observations.

2. Theory of the Observation

Let us suppose that the observed coronal emission arises in clots of subtelescopic sizes with a characteristic density $n_e \approx 10^9$-10^{10} cm^{-3}, and in the flares with a density 10^{12} - 10^{15} cm^{-3}. The number of this clots decreases, outward bound giving rise to the fall-off in intensity [4]. Considering all solar formations characterized by subtelescopic inhomogeneities,suggests that the corona is inhomogeneous on a subtelescopic level too. To determine which model corresponds more closely to reality, the white light image must be supplemented by one in the Balmer continuum ($\lambda < 365$ nm). The

Z. Mouradian and M. Stavinschi (eds.),
Theoretical and Observational Problems Related to Solar Eclipses, 177-180.
© 1997 Kluwer Academic Publishers.

emission intensity fall-off law in this filtergram reflects the decrease of clot numbers with radial distance (T_e and n_e are assumed to be equal for all clots).

If the fall off in emission intensity is simply caused by a decrease of electron density, the image (flare or prominence) the continuous spectrum would than decline (or in one of its bands). Plasma emission intensity from a unit volume in the Balmer continuum occurs while at the of photorecombinations and is proportional to $n_e^2 \, T_e^{-3/2}$, while the Thompson emission intensity is proportional to $n_e \, I_\lambda^\odot(r)$. Comparing the emission intensity of coronal plasma in a unit volume case with the corresponding intensity in the inner corona, the intensity of the Balmer continuum will appear to fall off approximately 10^4 - 3×10^5 times over a distance of two to three solar radii. For the same distances, the Thompson emission intensity falls correspondingly by 10^3 - 8×10^3. The difference in emission intensity between two corona images would be even more, since there is an integration of the emission along the line of sight. The observational intensity of Thompson scattering in each of the images (at a distance r from the Sun) obtained during eclipse is determined by the following integral along the line of sight:

$$I_{\lambda 1}(r) = 5 \cdot 10^{-11} \int_0^\infty n_e(l) I_{\lambda 1}^\odot(l) dl$$

The magnitude of the intensity of the photorecombinational radiation is:

$$I_{\lambda 2}(r) = 4.3 \cdot 10^{-29} \int_0^\infty n_e^2(l) T_e^{-3/2}(l) dl$$

So if the electron density in the solar corona, or in the other emissive objects, decreases monotonously with distance from the Sun, the image of the observed object in the Balmer continuum will have a greater intensity gradient than the image in one of the bands of the continuous spectrum (white light). For a structurally inhomogeneous object, the intensity distribution of its two images, which are determined both by clear Thompson scattering and by photorecombinations of the second quantum hydrogen level, the intensity gradients will differ in the opposite direction. The intensity gradients at a given radial distance from the Sun will fall in the continuous spectrum steeper than it will in the Balmer continuum. This is caused by the falling intensity, in the Balmer continuum at a distance from the Sun, that is proportional to the clot numbers along the line-of-sight. In the continuous spectrum, the fall of the emission intensity will happen both at the expense of a decrease in clot number, and at the expense of the dilution factor, with the distance of these clots from the Sun.

Emission intensity in the ultraviolet band ($\lambda < 365$ nm) is determined both by photo-recombination radiation and by the electron scattering. The contribution of the second component to the global intensity in the ultraviolet band can be calculated at each point in the image and depends on the mean intensity received in one of the bands of the continuous spectrum. The latter depends only on the Thompson scattering intensity $I_\lambda^\odot$. This procedure is correct since, having measured the intensity in one wavelength, we can be certain that at the same coronal point in another wavelength the

Thompson emission intensity will change proportional to solar emission intensity, because

$$I_{\lambda 1}(r) = I_{\lambda 2}(r) \cdot I_{\lambda 1}^{\odot} / I_{\lambda 2}^{\odot}$$

From this analysis, one can see that even a qualitative comparison of two filtergrams of the corona or solar object will allow us to answer the question as to which of the two the models represents the reality better. If the intensity decreases with distance from the Sun more smoothly in the Balmer continuum than in the continuous spectrum of visible bands, then it follows that the object has a subtelescopical structure and the microclots of higher density play a more important part than the monotonic decreasing electron concentration. If a gradient of the object intensity is steeper in the Balmer continuum than in the continuous spectrum, it means that subtelescopic inhomogenities play only a small part in the emission. If the intensity gradients on two photographs are identical, it means that structural inhomogeneities of subtelescopic sizes do have no realeffect on its emission, which is determined by the Thompson scattering in both the spectral bands.

3. Proposed Observations

Note that not only is the emission intensity in the Balmer continuum proportional to the emission measure,but line emission too. A coronal image in the Balmer continuum will differ from a line emission image. However, the images obtained in the Balmer continuum and in coronal lines differ significantly. This is caused by emission of the first type from relatively "cold" regions of corona, and of the second type from the "hot" regions. Even emission in the L_{α} line, which represents neutral hydrogen (that is in the "cold" region) will differ from the emission in the Balmer continuum, which is also determined mainly by relatively cold regions. This depends on the fact that the emission intensity decrease of the object in the line L_{α} is widely connected with the resonant scattering of the Sun's radiation, which means it depends a dilution coefficient, whereas photorecombination emission in the Balmer continuum depends on this coefficient to a lesser degree.

We need two filtergrams: one in the Balmer continuum (λ <365 nm), with the bandwidth of 20-50 nm, the other in a visual region chosen to exclude bright emissive lines. These could be, for example, spectral bands in $\lambda = 460 \pm 40$ nm or $\lambda = 580 \pm 50$ nm.

It would be useful to have filtergrams in the hydrogen or helium continuum as well. Filtergrams in L_c-continuum would be especially valuable as a diagnostic of photorecombination in "cold" corona regions, where this process exceeds the emission in the Balmer continuum by more than two orders. At the same time, the background intensity in the band of the Balmer continuum is greater than in the L_c-continuum by two orders. This would be a favorable condition for exploring subtelescopic inhomogeneities from observations of the L_c-continuum. This way, space-based observations may advance our knowledge of coronal structure.

References

1. Baumbach S. , (1937), *Astr. Nachrichten*, **263**, 121.
2. Rusin V., Rubansky M.:, (1990), *Slnechna Korona*, Sloven. Acad. Vied,Bratislava, p.157.
3. Shklovsky I.S., (1952), *Solnechnaya Corona*, Moskow, p.487.
4. Yakovkin N.A.:, (1991), *"Cold" Solar Corona*, Preprint, Academy of Sciences of the Ukrainian SSR, ITP-91-37E, p.28.

ECLIPSE OBSERVATIONS RELEVANT TO THE CORONAL HEATING PROBLEM

J M. PASACHOFF
Williams College--Hopkins Observatory,
Williamstown, MA 01267,   USA

Abstract. We are studying the coronal oscillation spectrum and its implications on the heating of the solar corona through the reduction of our data from observations of the total solar eclipse of 24 October 1995 and with further theoretical and observational investigations. The observations provide tests of proposed mechanisms to explain the heating of the solar corona via weakly compressive magnetohydrodynamic waves. We observed part of the lower corona in the [Fe XIV] green line with a CCD imaging at 5 frames per second from our eclipse site in Mukandgarh, Rajasthan, India. The new observations improve on our previously published observations in a variety of ways, especially by providing a 200 x 300 arcsec field of view instead of only a single point, which at this and at the 1998 and 1999 eclipses should allow us to explore the k-omega space of the coronal oscillation spectrum and thus enable us to distinguish between standing and propagating waves. We are also mapping the electron temperature of the corona through study of the broadening by electron scattering of ultraviolet Fraunhofer lines.

1. Introduction

The coronal heating problem has been the subject of numerous review papers, including Kuperus, Ionson, and Spicer [17], Hollweg [8]; Narain and Ulmschneider [20, 21], Gomez [3], Ulmschneider, Priest, and Rosner [36]; and Zirker [39]; see also Golub and Pasachoff [2]. Our experiment extends coronal observations into a new temporal domain, since nobody previously has observed coronal loops in emission lines with good statistics and with time resolution better than about 2 s (15 s in visible wavelengths). Historically, in astronomy, major discoveries have extended our science into previously unknown directions when new temporal or other domains have been opened; the discovery of pulsars, for example, came about when new apparatus had a much shorter time constant than previous apparatus. So, without claiming that such major discoveries will necessarily follow, the opening of the short-period time domain in coronal loops has the potential of discovery, adding further possibilities to the specific goals of our proposal.

Our original motivation was to test, in particular, the surface-Alfvén-wave model of Ionson [13, 14, 15]. The observations are also sensitive enough to test later calculations by Hollweg and collaborators [5, 6, 7], Hollweg, Jackson, and Galloway, [10], see also Ofman, Davila, and Shimizu [22]. These models propose that the Sun can pump energy into loops preferentially at certain resonant frequencies. The resonant frequencies are determined by the condition that an integral number of half-wavelengths fit into the length of the loop. Short loops, in particular, may have

181

Z. Mouradian and M. Stavinschi (eds.),
Theoretical and Observational Problems Related to Solar Eclipses, 181-187.
©1997 *Kluwer Academic Publishers.*

resonant frequencies in the range in which our instruments are sensitive. The detection of high-frequency oscillations has other ramifications as well. These waves may be particularly subject to dissipation via electron heat conduction or compressive viscosity [4, 34, 35]. Dissipation is, of course, necessary for heating, and the mechanism by which surface Alfvén waves and their relatives dissipate is a subject of ongoing discussion. Moreover, McKenzie, Banaszkiewicz, and Axford [18] have postulated the existence of high-frequency waves in the 0.01 Hz -10 kHz range. They were particularly concerned with the heating of the high-speed wind in coronal holes, where high-frequency waves could heat protons and heavy ions through pitch-angle scattering induced by the ion cyclotron resonance. Although we cannot probe the cyclotron resonant frequency, the detection of waves at the high-frequency limit of our experiment would lend at least conceptual support to this hypothesis.

2. Past Observations

I first carried out an early version of the experiment in collaboration with Donald A. Landman, then of the University of Hawaii, at the NSF site in India at the 1980 total solar eclipse, with time resolution of 0.1 second [31]. The results, summarized below, made it worthwhile to do a more sophisticated version, which we carried out at the NSF site in Indonesia at the 1983 total solar eclipse. Again we found a positive effect [30]. What we learned on these occasions suggested modifications for our equipment. The combination of our prior results Q with 3.5 minutes of observations in a clear sky the first time and 5 minutes in a cirrus-covered sky the second time Q with the results we hope to extract from the recent 1995 eclipse observations and from observations at the 1998 and 1999 eclipses have the potential of testing some of the many remaining theories of coronal heating based on wave dissipation. The 1984 eclipse in Papua New Guinea was too short and too soon for funding support, so we did not do a major experiment at it, though we were able to reduce and report auxiliary data relevant to the solar diameter and to the structure of coronal streamers [32, 33]. The fact that the weather at that eclipse was the clearest until the similarly short 1995 Indian eclipse indicates that it is important that experiments be carried out at all possible eclipses.

Our oscillation-experiment equipment was satisfactorily set up on site in Hawaii for the 1991 eclipse to extend our observations [23, 24, 25], but cloudy weather prevented us from observing the corona [26, 29]. Mainly higher sites in Hawaii escaped the unusual storm (see, for example, [16, 38], which includes a discussion of various experiments made at eclipses). At the 1994 eclipse in Chile, we changed the apparatus to use CCDs instead of fiber-optics and photomultipliers, but were troubled by clouds. We digitized at 10 Hz. At the 1995 eclipse in India, though totality was much briefer, we obtained data, digitized at 5 Hz, whose reduction we are now carrying out.

We were also able to carry out in Chile and India a coronal-temperature-mapping experiment, using a slow-scan CCD that was already part of Williams College's equipment. This experiment was based on photometric comparison of the corona at different ultraviolet wavelengths chosen for their sensitivity to thermal smearing of the solar spectrum reflected by coronal electrons.

3. The Williams College Coronal-Oscillation Experiment

The basic experiment is to observe the corona in two different wavelengths in order to detect coronal intensity fluctuations while eliminating terrestrial atmospheric effects. The main optical beam from the telescope is divided into two separate channels, thus far with a dichroic beamsplitter. One beam then passes through a DayStar [Fe XIV] 5303-Å, temperature-tuned interference filter, with a 3 Å FWHM passband. The other beam goes through a 100Å-wide continuum filter at a nearby wavelength band that has been selected to be as free as possible of other known coronal emission lines. After passing through the filters, the beams are imaged on a thermoelectrically cooled Princeton Instruments CCD. The image scale is approximately 2.0 x 2.0 square arcseconds per pixel, with a total field of view of 3 x 5 square arcminutes. The image data are digitized and recorded on hard disk. The specific CCD detector was selected to avoid any spurious effects from chip nonuniformities, especially from sub-pixel quantum efficiency variations. We binned 2 x 2 to minimize quantum-efficiency variations.

We observed the 3 November 1994 eclipse through a thin layer of clouds from the International Astronomical Union site at Putre, Chile, for 2 minutes 55 seconds, digitizing at 430 kHz with 16-bit accuracy. Images of the two beams were taken at a 10 Hz rate on separate Princeton Instrument CCD's. Data reduction reached only 3% accuracy, insufficient for the predicted effect, because of the low atmospheric transmission and the higher than typical sky brightness fluctuations from clouds. We reduced the data during the summer of 1994 [27, 28] using IRAF and IDL.

For the 24 October 1995 eclipse, we had no additional source of funds and could not again obtain the second Princeton Instruments CCD. (For 1994, one was purchased while the second was only on loan.) We thus built an optical train using corner prisms to bring the continuum-channel image onto the same CCD as the line image. Since we were acquiring twice as much information in each image, we were forced to reduce our exposure rate to 5 Hz for this experiment. In spite of our attention to baffling, internal reflections contribute to some cross-talk contamination between the two channels and to some internal reflections, the effects of which we must remove from the data. We are proposing a new optical train to eliminate this cross-talk at future eclipses. For the 1995 eclipse, weather prospects were very poor at the points where the eclipse would be longest, and so we observed from a site in Rajasthan at which totality was much shorter. This shorter totality coupled with a problem in computer cycling gave us 10 seconds of data in each of our channels, sufficient for searching for oscillations in the 1-Hz range. We used a telescope three times greater in area and had better atmospheric transmission properties than the year before. This factor of ten improvement in photon flux will increase our statistical accuracy by a factor of three and should approach the 1% level predicted to be necessary.

Our preliminary view of the 1995 data showed peak count rates of better than 4000 adc/bin in the coronal line spectrum, corresponding to more than 14,000 photons/bin. This value is in reasonably good agreement with expectations based on the factor of 3 from the larger objective, 2 from the 5-Hz exposure rates, and 2P3 from clearer skies compared with the Chile data. We thus achieved the 1% level we need. However, part of the gain resulted from slowing the sampling time to 5 Hz to capture both channels on the same CCD, using a larger part of the chip.

Our equipment for searching for coronal oscillations is much improved and much more sophisticated than the version used for the 1980 and 1983 experiments, which led to our previous published results, in which single points of the corona were picked out of the focal plane with bent-end fiber optics and fed through filters to cooled photomultipliers with much more rudimentary digitization. In our post-processing of the data we can align all the images very precisely using the lunar limb, to a precision better than one arcsecond. This procedure allows us to remove any image motion produced by tracking errors or atmospherically induced image motion, whereas the earlier versions of the experiment were completely dependent on tracking and atmospheric stability. The 1994 eclipse data showed the effect of periodic error in the drive, which we not only could remove but which also led to the choice of a telescope with a different drive for the 1995 eclipse. In addition, the imaging capabilities give us a 10,000-fold increase in the number of data points observed and allow us to isolate coronal loops in our search for oscillations. Finally, using images, we can look for decreases with height above the limb of the power of any detected oscillations, possible evidence of propagation and dissipation.

We found indications of excess Fourier power near 1 Hz in the 1980 experiment [31] and again found excess power in the 1983 observations [30]. Our conclusions were restrained, given that we had data from single-channels: our results "are consistent with theories of fast mode waves in a magnetically dominated regime," but that "Further observations are necessary to completely confirm and quantify these coronal oscillations," to assess atmospheric and other contributions." We are pleased that we can now be carrying out this next-generation CCD set of observations.

4. The Coronal Temperature Map

In addition to our coronal oscillation experiment, we also carried out an experiment using the Photometrics CCD that we already had at our observatory to map the coronal electron temperature.

I have been interested in using the effect of coronal scattering on Fraunhofer lines since Menzel and Pasachoff [19], when we showed that any such dip was within the calibration uncertainty of the films that had been used for the observations. The more recent use of our 512 x 512-pixel CCD in 1994 and 1995 allowed simultaneous measurements to be taken at many positions in the corona, not only showing possible temperature variations but also improving the statistical accuracy of measurements for averaged areas. When the experiment is successfully conducted under clear skies, the results should be comparable to coronal electron temperatures deduced by rocket or satellite observations of the Lyman-alpha broadening, which determine electron temperatures as a secondary effect to proton temperatures [37] and from space, especially with SOHO's UVCS.

This method is based on theoretical calculations of Cram [1]. The method seeks to see if the many ultraviolet spectral lines in the spectrum of the solar photosphere, when Doppler-broadened by coronal electrons, cause a slight dip in the resulting continuous spectrum of the corona. Cram calculated the thermal Doppler broadening of the solar spectral features that are Thomson-scattered by the coronal electrons at the high coronal temperatures. While most solar spectral lines are completely eliminated by the broadening function, which has a width around 100 Å, the more prominent features, such as the calcium-II H and K lines, and the G-band, do show up as broad dips of a few per cent in the scattered spectrum. The higher the coronal electron temperature, the greater the thermal broadening and the shallower these residual dips. Cram has shown that accurate measurement of the intensities at the wavelengths of these residual features would lead to a determination of the coronal electron temperature with minimal assumptions (which is not the case with most other methods of coronal temperature measurement). Since the corona is not in equilibrium, determination of the coronal temperature by different methods is important for understanding coronal excitation and structure.

Cram's calculations showed that by determining the intensities at several wavelengths near 3900 Å, near the calcium lines, to an accuracy of 1%, the coronal temperature will be determined to the relative high accuracy of 200,000 K. Reardon, then at the Institute for Astronomy of the University of Hawaii, recalculated the scattering problem for us using finer wavelength resolution and more capable computers, and concluded that we should observe at slightly different wavelengths, which we did. Cram had calculated that the ratio of the hottest and the coldest temperature values considered would be maximum at 3900 Å and its inverse would be maximum at 4100 Å. The ratio should be unity at 4000 Å, so we planned an additional measurement at this node as a normalization. Reardon's modifications gave 3875 Å, 4050 Å, and 4410 Å as the more desirable wavelengths to observe, and we had new filters made with relatively square passbands of about 40 Å. Our ultraviolet set of filters avoided the coronal emission lines, especially [Fe XI] 3987 Å, [Ca XII] 4086 Å, and [Ni XII] 4231 Å. We also made the necessary polarization measurements to isolate the K- and F- coronal components at each of our wavelengths. IchimotoUs group from the Japanese National Astronomical Observatory made similar observations in Chile but with wider filters with a passband that is less square. Ichimoto has thus far reduced his spectral data on this subject [11, 12] but cannot reduce the CCD imaging through filters because of the nonphotometric sky. He confirmed that the intensity was reduced by a factor of a few by clouds in Chile at the 1984 eclipse; he did not repeat his experiment in 1995. Our calculations show that our choice of filters should be more sensitive to the temperature than his. Ichimoto has shown that the method also has capabilities of measuring solar-wind velocities.

Though the skies in Chile for the 1994 eclipse were not of photometric quality, we could assess the equipment and data and refine our required calibrations.

The briefness of totality in India led to a compression of our observing sequence--limiting the polarization sequences, in particular--but the data we obtained seem good and we are analyzing them. The longer totality available in 1998 and 1999 will lead to our mapping extending farther above the solar limb.

The recent report from the Solar and Heliospheric Observatory's UVCS experiment of oxygen coronal temperatures of 100 MK show the importance of carrying out as many independent assessments of temperature in this extremely non-equilibrium solar corona.

Acknowledgments.

Bryce Babcock of Williams College supervised the oscillation experiment, both in Williamstown and on site. Kevin Reardon, now of the Osservatorio di Capodimonte, Naples, participated in all aspects of the experiments. We also thank Jonathan Kern of New Orleans, who made the optical train for the 1995 version, and Eric Kutner, then of Princeton University. Lee Hawkins of Wellesley College, coordinator for the Keck Northeast Astronomy Consortium, ran the temperature experiment on site at the 1994 and 1995 eclipses. Zadig Mouradian of l'Observatoire de Paris at Meudon participated on site in Chile. Williams College students Bonnie Schulkin and Robert Galloway worked on site at the 1994 eclipse and Williams College student Sebastian Diaz and Keck-sponsored students Rana Nichols-Kiley and David Berger of Colgate University and Matthew Pickard of Vassar College worked on site at the 1995 eclipse and/or on the data in Williamstown.

This research has been supported by the National Science Foundation's Atmospheric Sciences Division, Astronomy Division, and Education Division, most recently through grants ATM-9005194, AST-9014889, USE-9050643, ATM-9207110, DUE-9351279, AST-9512216, and by grants from the Committee on Research and Exploration of the National Geographic Society. Our expeditions from Williams College, and student participation, have also benefitted by grants from the W. M. Keck Foundation and from the Brandi Fund and the Safford Fund.

References

[1]. Cram, L. E., (1976), *Solar Physics* **48**, 3-19.

[2]. Golub, L., and J. M. Pasachoff, (1997), *The Solar Corona* ,Cambridge University Press.

[3]. Gomez, D. O., (1991), *Fundamentals of Cosmic Physics* **14**, 131-233.

[4]. Gordon, B. E., and J. V. Hollweg, (1983), Astrophys. J. **266**, 373-382.

[5]. Hollweg, J. V., (1981), *Solar Phys.* **70**, 25-66.

[6]. Hollweg, J. V., (1984a), *Astrophys. J.* **277**, 392-403.

[7]. Hollweg, J. V., (1984b), *Solar Phys.* **91**, 269-288.

[8]. Hollweg, J. V., (1990), *Computer Phys. Reports* **12**, 205-232.

[9]. Hollweg, J. V., (1991), in *Mechanisms of Chromospheric and Coronal Heating*, (ed) P. Ulmschneider, E.R. Priest and R. Rosner, 423-434, Springer Verlag.

[10]. Hollweg, J. V., S. Jackson, and D. Galloway, (1982), *Solar Phys.* **75**, 35-61.

[11]. Ichimoto, K., K. Kumagai, I. Sano, T. Kobiki, T. Sakurai, and M.Alberto, 1995, *Proceedings of the International Symposium on the Total Solar Eclipse of November 3, 1994*, Lake Titicaca, M. S. Raljevic, F.Zaratti, and J. M. Pasachoff, (eds)., 72-87.

[12]. Ichimoto, K., K. Kumagai, I. Sano, T. Kobiki, and T. Sakurai, and A.Munoz, (1996) *Pub. Astron. Soc. Japan* (in press).

[13]. Ionson, J. A., (1977) Ph. D. Thesis, Univ. Maryland.

[14]. Ionson, J. A., (1978) *Astrophys. J* ,**226**, 650-673.

[15]. Ionson, J. A., (1979), private communication.

[16]. Koutchmy, S., and M. Molodensky, (1992) *Nature* **360**, 717-719.

[17]. Kuperus, M. , J. A Ionson, and D.S.Spicer, (1981) *Ann. Rev. Astron. Astrophy.*, **19**, 7-40.

[18]. McKenzie, J.F., M.Banaszkiewicz, and W.I.Axford, (1995), *Astron. Astrophys*, **303**, L45-L48.

[19]. Menzel, D.H., and J. M. Pasachoff, (1968) *Publ. Ast. Soc. Pacific*, **80**, 458-461.

[20]. Narain,U., and P.Ulmscheider, (1990) *Space Science Reviews* , **54**, 377-445.

[21]. Narain,U., and P.Ulmscheider, (1996) *Space Science Reviews*,**75**, 453.

[22]. Ofman,L., J.M. Davila, and T. Shimizu, (1996), *Astrophys. J.(Lett.)*. **459**, L39-42.

[23]. Pasachoff, J. M., (1991a) in *Mechanisms of Chromospheric and Coronal Heating*, P. Ulmscheider, E.P.Prist, R.Rosner, (eds.), Spriger-Verlag, 1991, 25-29.

[24]. Pasachoff, J. M., (1991b) in IAU Colloquium #130 on *The Sun and Cool Stars*, I. Tuominen, D. Moss, and G.Rüdinger (eds.) Spriger-Verlag, 1991, 285-287.

[25]. Pasachoff, J. M., (1991c), in *Flare Physics in Solar Activity Maximum 22*, (1991), Y. Uchida et al., (eds.), Springer-Verlag,. 283-287.

[26]. Pasachoff, J. M., (1992), in *National Geographic* **181**, May, #5, 36-37.

[27]. Pasachoff, Jay M., B. Babcock, and K. P. Reardon, (1995), *Proceedings of the International Symposium on the Total Solar Eclipse of November 3,1994*, M. S. Raljevic, F. Zaratti, and J. M. Pasachoff,(eds.), Revista de la Academia Nacional de Ciencias de Bolivia No. **69**, 18-21.

[28]. Pasachoff, Jay M., B. Babcock, J. S. Diaz, K. P. Reardon, and R.Nichols-Kiley, (1996), *Bull. Am. Astron. Soc.* G, No. 4, 1995, 1427.

[29]. Pasachoff, Jay M., and Michael Covington, (1993), *The Cambridge Eclipse Photography Guide*, Cambridge University Press.

[30]. Pasachoff, J. M., and N. F. Ladd, (1987), *Solar Phys.* **109**, 365-372.

[31]. Pasachoff, J. M., and D. A. Landman, (1984), *Solar Phys.* **90**, 325-330.

[32]. Pasachoff, J. M., and B. O. Nelson, (1987), *Solar Phys.* **108**, 191-194.

[33]. Pasachoff, J. M., K. P. Reardon, and J. MacKenty, (1993), *Solar Phys*.**146**, 405-408.

[34]. Porter, L. J., J. A. Klimchuk, and P. A. Sturrock, (1994a), *Astrophys.J.* **435**, 482-501.

[35]. Porter, L.J., Klimchuk, J. A., and Sturrock, P. A., (1994b), *Astrophys.J.* **435**, 502-514.

[36]. Ulmschneider, P., E. R. Priest, and R. Rosner, eds. (1991), *The Heating of the Solar Chromosphere and Corona*, Springer-Verlag.

[37]. Withbroe, G., J. L. Kohl, H. Weiser, and R. H. Munro, (1982), *SpaceScience Reviews* **33**, 17-52.

[38]. Zirker, J. B., (1981/1995), *Total Eclipses of the Sun*, 2nd ed.,Princeton Univ. Press., 1981, reprinted 1995 with a postscript.

[39]. Zirker, J. B., (1993), *Solar Phys.* **148**, 43-60.

THE INDICATION OF NEUTRAL HYDROGEN IN THE SOLAR CORONA

I. DOROTOVIC, B. LUKÁC
Slovak Central Observatory, P.O. Box 42, SK-94701 Hurbanovo,
SLOVAKIA

M. MINAROVJECH, M. RYBANSKÝ
Astronomical Institute of the Slovak Academy of Sciences, SK-05960
Tatranská Lomnica, SLOVAKIA

Abstract. This paper describes an experiment carried out during an eclipse to determine the colour of the solar corona. It contains the results of the solar corona colour measurements during the total eclipse of 1994 near the solar limb. The difference in the colour of the corona and the colour of the photosphere might indicate the presence of neutral matter. Moreover, it theoretically indicates the colour according to the present model of the corona. On the basis of the results, it seems that solar corona should be redder everywhere than the solar disc centre even near the limb. The experiment shows that the corona near the limb is bluer than the centre of the solar disc, and this effect decreases with the height in the corona. We compared the results of the experiment with those of the theoretical calculation of the colour. This comparison shows that the experimental results are in agreement with the theoretical, above the height of 3'. This fact probably proves the presence of neutral hydrogen in the corona.

1. Introduction

Theory is usually based on the interpretation of observations and measurements. It is possible that the experimental data used to determine the parameters of coronal matterhave been wrongly interpreted. Present coronal heating theory is based on a magnetohydrodynamic approach, which assumes that the matter in the corona is entierly ionizid and highly conductive. But if the ionization were not complete, i.e. if there is also some neutral hydrogen in the coronal matter, its conductivity would rapidly decrease. This fact could lead to a revision of MHD model of the solar corona and the adoption of a new approach that might us understand the coronal heating better, as well as the transport of the matter to the corona. The presence of neutral matter may be indicated by a change inthe colour of the corona, because in that case it would be due to Rayleigh rather then Thomson scattering,which would hold in the case of complete ionization.

The latest experiments conducted to estimate the colour of the corona led to the conclusion [5] that within the precision of measurements, this colour is the same as that of the photosphere, as far as the inner corona is concerned, and the proportion of red increases with distance, because of the F-corona contribution. All papers dealing with colour determination from the colour photographs taken during the eclipse [1, 4] are also based on this assumption In contrast to this authers, we study the soundness of

189

Z. Mouradian and M. Stavinschi (eds.),
Theoretical and Observational Problems Related to Solar Eclipses, 189-193.
©1997 *Kluwer Academic Publishers.*

this assumption, both theoretically and experimentally. Presently, the precision of corona color measurements using CCD cameras is at least one order better than with photographic photometry. In our opinion, the question of the solar coronas color may soon be solved with CCD cameras.

This paper describes an experiment to determine the colour of the corona during an eclipse, and compares the experimental datd with the theoretical model. We observed the total solar eclipse on November 3, 1994. The instruments used, and the whole observation are described elsewhere [3], hereafter Paper I.

2. Method of Reduction and Results

The solar corona colour is defined as the ratio of the spectral intensity between the blue and the red part of the spectrum. We define this ratio with respect to the photospheric spectrum. A test of the "flat field" of the camera showed that the photospheric spectrum measurements in different rows are within the limits of precision corresponding to the theoretical precision of measurements. It then becomes possible to create only one averaged record from the photospheric spectrum. The data reduction method was described in Paper I.

Values less than 10 (the maximum level being 256) cannot be used for our purposes in order for the error in the intensity ratio not to exeed 5%. The data from the records Nos. 3 - 5 (Paper I) that fullfil both criteria were divided into three sets according to the height of measurement in the solar corona (3, 5 and 7 arcmin.). The final corona/photosphere ratio is shown in Figure 1. According to this Figure the solar corona seems to be getting redder the closer we get to the solar surface (in the mentioned heights).

The above method cannot be used for measurements in the inner solar corona, within approximately 3'. In order analyze this part of our measurement, we need to derive a spread function which, in our case, will depend not only on the height in the solar corona but also on the wavelength. If we know the spread function, a deconvolution of it will enable us to use the data from the inner solar corona.

To eliminate the scattering near the limb, a stray light function has to be determined. According to Zwaan [7], this is how we indicate total light, i.e. what is in the picture but not radiated by the object. We use the calibration measurement of the solar disk spectrum for determing this function. Figure 2 shows the resulting spectral intensity ratio in the red and blue parts of the coronal spectrum.

In spite of a rather low precision estimated at about 2 - 3%, Figure 2 shows that the colour of the corona is bluer than that of the solar disk centre.

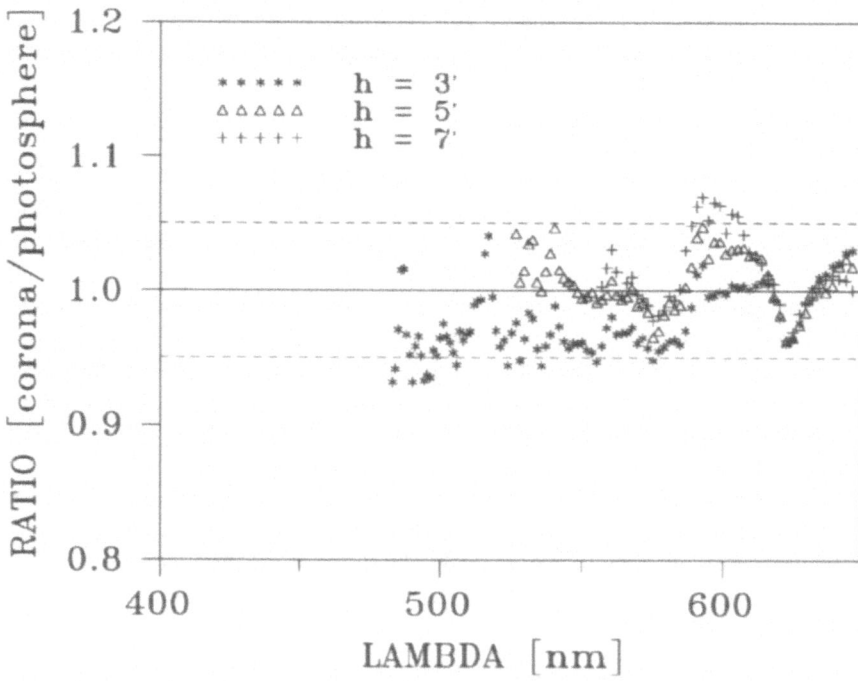

Figure 1. The ratio of spectral intensities $I_{corona}/I_{photosphere}$ for the intermediate heights in the corona (for details see the text).

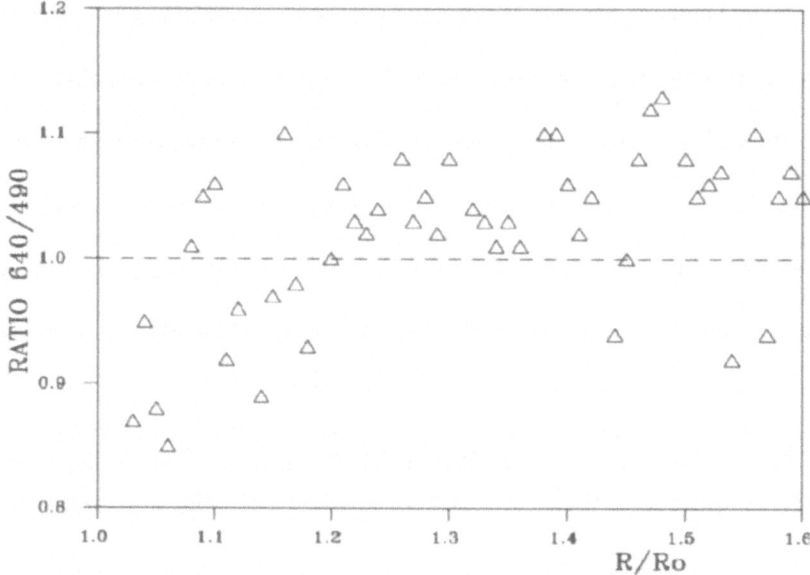

Figure 2. The spectral intensity ratio I_{640}/I_{490} (at the centre of the solar disc I_{640}/I_{490} = 1) as a function of a distance from the solar centre.

If we attributethe difference to scattering on neutral hydrogen atoms there would have to be about 7 of them for each free electron. So the degree of ionization in the inner corona would be relatively low.

3. Theoretical Model for the Solar Corona Colour

The colour of the corona depends on the scattering conditions of the radiation coming from the photosphere, and yields information concerning the radiation mechanisms in the corona. We assume that the integral spectral intensity of the corona is caused by a contribution from photospheric radiation scattering on free electrons in the K-corona (Thomson scattering) and by a contribution from photospheric radiation scattering on dust particles in the F-corona. If neutral hydrogen is present in the corona, the inner corona would become bluer because of Rayleigh scattering. The complete calculation of scattering in white corona has been described elsewhere [2]. The colour, in this paper is defined as the ratio of the spectral intensities at 640 and 490 nm.

On the basis of these results, it seems that the solar corona should be redder than the solar disk centre everywhere, even near the limb. With increasing height, this "redness" first decreases and then increases under the influence of the F-corona.

4. Discussion and Conclusions

The measurements show that the colour of the corona is in agreement with the intensities calculated for heights above 3' from the limb. At lower heights, the corona is bluer, which might be caused by the presence of neutral hydrogen. However, this interpretation is problematic because the stray light function, obtained by Fourier transform near the limb, is not unambiguous.

If the ratio of hydrogen atoms to free electrons is equal to one, then the spectral intensity ratio changes by 0.12% [6]. This means that the precision of colour measurements should be at least on the same level. Current CCD cameras with 16-bit converters make such precision possible.

We are currently investigating results from a similar experiment performed at two observing stations (India, Neem Ka Thana and Thailand, Nakhon Sawan) during the eclipse of October 24, 1995. ST-7 CCD cameras with 16-bit converters were used in these experiments.

We would appreciate if the experiment could be repeated by other expeditions during future eclipses, if they have better equipment for observation.

Acknowledgements. The expedition was made possible thanks to the decisive financial and organizational support of the Ministry of Culture of the Slovak Republic. This work was supported by the GAV Grant 2004/95 of the Slovak Academy of Sciences.

References

1. Alexeyeva, I.V., Kroussanova, N.L., Streltsova, M.V.: (1994), in *IAU Coll. 144, Solar Coronal Structures,* (ed). V.Rušin, P.Heinzel and J.-C.Vial, VEDA, Bratislava, 597-599.
2. Dorotovic, I., Rybanský, M.: (1996), *Solar Phys.*,(in press).
3. Lukác, B., Minarovjech, M., Rybanský, M.: (1995), in *Proceedings of the Int. Symp. on the Total Solar Eclipse of November 3, 1994,* Acad. Nac. Bolivia 69 , La Paz, 111-117.
4. Nikolsky,G.M., Nesmyanovich,I.A.: (1983), *Astron. Zh.,* **60** , 1179 -1186.
5. Shklovsky,I.S.: (1962), *Fizika Solnecnoy Korony* , GIFML, Moscow.
6. Unsöld,A.: (1968), *Physik der Sternatmosphaeren* , Springer Verlag, Berlin-Heidelberg-New York.
7. Zwaan,C.: (1965), *Recherches Astronomiques de l'Observatoire d'Utrecht,* XVII (4).

RELEVANT RADIO OBSERVATIONS DURING A TOTAL SOLAR ECLIPSE

A. ONCICA
Astronomical Institute of Romanian Academy
Str. Cutitul de Argint 5, 75212 Bucharest - Romania

1. Introduction

The opportunity of observing a total solar eclipse opens the path to several specific experiments. These observations can reveal more information on structures and phenomena belonging to the solar atmosphere. Fundamental for the optical part of the spectrum is the occultation of the overwhelming photospheric flux. During the few minutes between the second and third contact it is possible to observe the structures of the faint corona and (for few seconds) the edge-on chromosphere.

However, in the radio part of the spectrum, the radio flux from the solar photosphere is no longer dominant. This allows the investigation of the solar disk and the structures in the solar atmosphere throughout the eclipse. From the point of view of solar radio astronomy relevant observations may be carried out for several hours between the first and the fourth contact. For the same reason, the observations may also be carried out from locations near the path of totality.

The main point is that the time variation of the received radio flux is related to the spatial distribution of radio sources. As the Moon limb sweeps across the Sun, it translates the spatial information into the time domain. With suitable deconvolution techniques, supported by good time resolution and spectral information, the 3D structure of the radio sources can be extracted[1], [2], [3].

This paper attempts to summarize some of the relevant solar radio observations that can be carried out during a (total) solar eclipse based on previous experience and present knowledge of solar atmospheric structures as seen in the radio spectrum.

2. Radio Observations

All the information a radio telescope can collect is contained in the spatial, spectral and time distribution of the radio flux intensity(in both polarizations):

$$I = I_{\pm}(\theta, \phi, h, t, f) \tag{1}$$

Here, θ and ϕ are to be regarded merely as coordinates on the solar disk, and h as the height above the photosphere. The 3D information reveals the geometry of the source.

195

Z. Mouradian and M. Stavinschi (eds.),
Theoretical and Observational Problems Related to Solar Eclipses, 195-198.
©1997 *Kluwer Academic Publishers.*

Adding the polarization, the geometry of the magnetic field can be inferred. The spectral content yields information concerning physical parameters such as electron density and electron temperature. Unfortunately, no such versatile instrument exist. Solar radio telescopes, like any other instrument, are designed and built for maximum output but only for a limited range of information [4].

Radio polarimeters record only the time variation at a fixed frequency, over a fixed frequency bandwidth, and through a fixed antenna beamwidth. Increased time resolution (to resolve rapid phenomena) and increased frequency resolution (not to smear out spectral details) are obtained at the expense of sensitivity. Increased spatial resolution (pencil beam) is obtained at the expense of time resolution, as mapping is performed one pixel at a time.

Radio spectrographs can provide both high time resolution, and high spectral resolution but not necessarily both in the same time. As they are designed for the surveillance of the whole solar atmosphere, they have no spatial resolution. But as we can anticipate certain spatial information, this is included in the height-above-photosphere dependence of the frequency.

Radio imaging instruments can obtain 2D information concerning the source geometry. Their spatial resolution depends on the quality of the Fourier plane coverage and the way it is obtained. Their time resolution is poor, due to the time spent in reconstructing the image. Their spectral resolution is also poor, due to the principle of interferometers operation. Sometimes the spectral information can be improved by using simultaneous multi-frequency observations.

3. Radio Diagnostics

Radio observations allow us to probe the solar atmosphere from the chromosphere to the corona. The frequency of the emitted radiation, whether at the plasma frequency or at the gyrofrequency or its harmonics, depends on the height above the photosphere, where was generated [4]. Different emission processes allow for different diagnostics. As the emission processes are frequency-dependent, it is of great importance to have good spectral coverage in order to generate a good diagnostic [4].

Thermal free-free emission, depending on whether it is optically thin or thick, provides diagnostics of electron density and/or temperature. EUV line observations yield an independent mean, and the dependence of each line differs in accordance with the temperature of the emission region. In the millimetric range, observations have been made to probe the chromosphere-corona transition [5], [6] and the filament-corona transition [7], [8] region. The information concerning the strength and topology of the longitudinal magnetic field is present in the degree of circular polarization.

Resonant emission processes may dominate the free-free radiation in strong magnetic fields. The spectral characteristics of this radiation depends on the energy spectrum of the radiating electrons and on the relation between the plasma frequency (electron density) and the gyrofrequency (the intensity of the magnetic field) [9].

Multi-frequency mapping potentially contains information on the magnetic field topology, electron density, and temperature as a function of height. High resolution observations were carried out at the VLA [10],[11]. At 15, 5 and 1.5 GHz, the resolution limit was 2, 6 and 15 arcseconds respectively. The claim was that no structures below 2.5" were detected. This may be the consequence of the "coronal seeing". This result must also related to the fact that the time-shared synthesis map of the three frequencies took 6 to 12 hours.

4. Eclipse Observations

The most important aspect of the radio observation of a (total) solar eclipse lies in the spatial resolution, which can be attained with only small antenna instruments. As the Moon's limb sweeps across the Sun at a rate of approximately 0.5 arcsec/s, one second integration time gives 0.5 arcsecond one-dimensional spatial resolution (at all frequencies). If the observation can be made at different frequencies simultaneously, then high spatial resolution spectroscopy of a particular source is possible. It is even better to use a radio spectrograph. A relevant example of the performance of such a procedure is the observation of the 7 March 1970 eclipse at Sagamore Hill Radio Observatory [12], on seven frequencies simultaneously.

One aspect that seems to have been forgotten is that the previous procedure assumes the validity of the laws of geometrical optics. But in the dm-cm range, the first Fresnel zone of the diffraction pattern of the Moon's limb is of the order of a few arcseconds. This means that sub-second time constant observations must take this phenomena into account. The technique of lunar occultation observation, developed for galactic and extragalactic radio sources, shows that the spatial resolution is limited only by the signal to noise ratio of the receiver [13].

Detailed calculations will show that today's solar radio spectrographs are well-suited to high spatial resolution observations during a (total) solar eclipse [3].

5. Conclusions

Apart from few dedicated experiments, most solar eclipse observations so far have been performed only in accordance with the accidental coincidence of the totality path and the location of the radio telescope. Optical instruments suitable for eclipse observation can be designed and built in a quite compact format so they can be easily transported to the eclipse site.

The contrary is true for radio telescopes and their heavy mechanical structures. Radio observations of solar eclipses have been carried out only when certain instruments happened to be at least near the path of totality. Exceptions exist only for the high-frequency part of the radio spectrum [2], [14], [15]. As the path of the August 11, 1999 total solar eclipse comes near most European solar radio instruments [16], this is a good opportunity for such observation.

Lastly, it is worth mentioning here that radio observations are much less sensitive to atmospheric conditions than optical observations are.

198

References

1. Hazard C. (1976) Lunar Occultation Measurements, in Meeks M. L. (ed.) *Methods of Experimental Physics*, Vol. 12, *Astrophysics*, Part C, *Radio Observations*, Academic   Press, New York, 92-117.
2. Hagen, J. P. et al. (1971) Observations of the 7 March, 1970 Total Solar Eclipse at Wavelengths of 3.2 mm and 8.3 mm, *Solar Phys.* **21**, 286-296.
3. Oncica A. (1996) A Radio Observation Project for the 1999 Total Solar Eclipse, (this proceedings).
4. Hurford G. J. (1992) Solar Radio Observations, in T. G. Schmelz and J. C. Brown, The Sun, A Laboratory for Astrophysics, Kluwer Academic Publishers, Dordrecht, 297- 312.
5. White, S. M. and Kundu, M. R. (1994) Observations of the 1991 Eclipse at 3.5 mm Wavelength, in D. M. Rabin, J. T. Jefferies and C. Lindsay (eds.) *Infrared Solar   Physics*, IAU Symp. No.154, Kluwer Academic Publishers, Dordrecht, 167-172.
6. Ewell Jr., M. W. et al. (1994) 850 μm Observations of the 11 July 1991 Total Solar Eclipse, in D. M. Rabin, J. T. Jefferies and C. Lindsay (eds.) *Infrared Solar Physics*, IAU Symp. No.154, Kluwer Academic Publishers, Dordrecht, 161-166.
7. Hiei, E. et al. (1986) Dark Filaments Observed at 8.3 mm and 3.1 mm Wavelength, in Poland A. I. (ed.) *Coronal and Prominence Plasmas*, NASA Conference Publications 2442.109-116.
8. Vial, J. C. (1986) Diagnostic and Structure, in Poland A. I. (ed.) *Coronal and Prominence Plasmas*, NASA Conference Publications 2442, 89-95.
9. Hurford, G. J. and Gary, D. E. (1989) The Owens Valley Solar Array, in R. M. Winglee and B. R. Dennis (eds.), Developments in Observations and Theory for Solar Cycle 22, MAX'91 Workshop #2, Laurel, Maryland, 328-332.
10. Kundu, M. R. (1986) VLA Observations of Solar Filaments at 6 and 20 cm Wavelength, in Poland A. I. (ed.) *Coronal and Prominence Plasmas*, NASA   Conference Publications 2442, 117-120.
11. Gary, D. E. (1986) The Microwave Structures of Quiescent Solar Filaments at High Resolution, in Poland A. (ed.) *Coronal and Prominence Plasmas*, NASA Conference Publications2442, 121-125.
12. Straka, R. M. (1971) Spectral Radio Observations of a Solar Eclipse, *Solar Phys.* **21**, 469-480.
13. Cohen, M. H. (1969) High-resolution Observation of Radio Sources, *Ann. Rev. of Astron. and Astrophys.* **7**, 619-664.
14. Nagnibeda, J. P. et al. (1993) Fine Structure and Temporal Variations of the Solar Radio Sources in mm Wavelength Range During the July 11, 1991 Solar Eclipse, in *Solar Coronal Structures, IAU Coll.No.144*, Sept. 20-24, Tatranska Lomnica,. VEDA Publ. 127.
15. Korhavin, A. N. (1993) Fine Coronal Structure of Solar Active Regions from Multi-frequency Microwave Observations of the July 11, 1991 Solar Eclipse in Mexico, in *Solar Coronal Structures, IAU Coll.No.144*, Sept. 20-24, Tatranska Lomnica, VEDA Publ. 128.
16. Pasachoff, J. M. and Covington, M. A. (1993) *The Cambridge Eclipse Photography Guide*, Cambridge University Press, Cambridge.

NASA BULLETIN FOR THE TOTAL SOLAR ECLIPSE OF 1999

F. ESPENAK
NASA/Goddard Space Flight Center
Planetary Systems Branch, Code 693
Greenbelt, Maryland 20771, U.S.A.

Abstract On 1999 August 11, a total eclipse of the Sun will be visible from parts of central Europe, Turkey, the Middle East and India. NASA is preparing a special bulletin on the eclipse which contains detailed predictions, tables, maps, and meteorological data useful for planning eclipse expeditions and scientific observations. Public information is also provided for eye safety, viewing and eclipse photography. The bulletin will be published by the end of 1996. It will also be available via the World Wide Web from: http://umbra.nascom.nasa.gov/eclipse/index.html.

1. Introduction

The total solar eclipse of 1999 August 11 is visible from heavily populated regions of central Europe. The path of the Moon's shadow begins in the North Atlantic and makes first landfall in southern England. Traveling eastward, the shadow crosses through parts of France, Belgium, Luxembourg, Germany, Austria, Slovenia, Hungary, Yugoslavia, Romania and Bulgaria (Figure 1). Local time of the total eclipse is mid morning to early afternoon, depending on the longitude. The center line duration of total eclipse gradually increases from 2m 10s in France to 2m 22s in Romania.

After crossing the Black Sea, the Moon's shadow sweeps through Turkey, Syria, Iraq, and Iran (Figure 2). In this region, total eclipse occurs in mid to late afternoon and the duration decreases from 2m 17s to 1m 26s. Continuing on its eastern course through Pakistan and central India, the path ends at sunset in the Bay of Bengal.

Although the total eclipse will be confined to a narrow corridor 55 to 112 km wide, a partial eclipse will be visible from a much larger region which encompasses all of Europe, northern Africa, the Middle East and the western half of Asia.

2. A Scientific and Educational Opportunity

Not since 1961 has the Continent witnessed a total eclipse of the Sun. This rare chance to observe the Sun's corona is already attracting a great deal of attention from both the international scientific and lay communities. The total eclipse not only affords solar astronomers with a unique opportunity to probe the Sun's atmosphere. It is also the most spectacular astronomical phenomenon visible to the unaided eye. As such, it provides an outstanding educational event to expand public awareness and

199

Z. Mouradian and M. Stavinschi (eds.),
Theoretical and Observational Problems Related to Solar Eclipses, 199-207.
©1997 *Kluwer Academic Publishers.*

understanding of astronomy, and to enhance interest in science at all levels from young children through senior citizens. To serve this diverse audience with accurate information about the coming event, NASA is preparing a special bulletin focusing on the 1999 solar eclipse.

Published through NASA's Reference Publication (RP) series, the eclipse bulletins contain detailed predictions, maps, and meteorological data for future solar eclipses of special interest. These publications are prepared in cooperation with the Working Group on Eclipses of the International Astronomical Union and are provided as a public service to scientists, educators, the general public and the media. Initiated in 1993, the bulletins are typically prepared 24 to 36 months before each major eclipse to allow adequate lead time for expedition planning purposes. The most recently published bulletin covers the total solar eclipse of 1998 February 26 (NASA RP1383).

To maximize the distribution of this information, the reproduction of any portions of the eclipse bulletins is strongly encouraged. This includes publication in newspapers and magazines or use by television or radio to inform the general public.

3. NASA Bulletin for the 1999 Total Solar Eclipse

The NASA eclipse bulletin "Predictions for the Total Solar Eclipse of 1999 August 11" Espenak and Anderson [1] is currently under preparation and will be published by the end of 1996. Copies of the 1999 bulletin will be distributed to the scientific community through a mailing list under development. Since the authors anticipate a large demand for information on this eclipse from the general public, a large number of additional copies will be printed and will be available upon request.

The basic contents of the 1999 eclipse bulletin include a brief, descriptive narrative, followed by a series of detailed figures, tables and maps of the eclipse path. The text describes the regions of visibility of both the penumbral and umbral shadows which are illustrated with maps. Examples of some of the more detailed maps show the path through Germany and Romania (Figures 3 and 4, respectively). Notable cities or locations in the path are highlighted along with umbral durations and solar altitudes. An explanation of the eclipse's Besselian elements and the shadow contact tables are followed by descriptions of various eclipse path and local circumstances tables (e.g.: Table 1). The text then describes the detailed umbral path maps appearing in the last section of each bulletin. These maps have a scale of 1:5,000,000, which is adequate for showing major cities, highways, airports, and basic topography required for eclipse expedition planning. Geographic coordinates are also provided to allow plotting the path on larger scale maps. A discussion is included which describes how to estimate the times of second and third contact for locations off the center line.

A fundamental parameter used in the prediction of solar eclipses is the Moon's mean radius "k". The text discusses how different values of "k" affect the predictions. The Moon's actual radius varies as a function of position angle and libration due to irregularities in the limb profile. This complicates the geometry of the umbral path at its northern and southern limits by creating narrow zones 5 to 10 km wide where a grazing eclipse will be seen. The bulletin describes on how to use the tables to plot these graze zones on high resolution maps. A discussion of the Saros history of the eclipse puts the event into a historical perspective.

One of the more important sections of the narrative focuses on the weather prospects along the eclipse track. This information is critical in the selection of a site whose meteorological characteristics favor clear skies during the eclipse. Appropriate climate statistics in the form of tables and figures lend support to this discussion.

Several sections provide information directed specifically at the lay audience. An expanded discussion on eye safety provides suggestions on how to view a solar eclipse without injury. The appearance of the sky during totality describes which bright planets or stars which may be visible to the naked eye. A concise set of instructions for eclipse photography attempts to optimize the layman's chances of successfully capturing this extraordinary event on film or video.

To maintain a high level of accuracy and insure consistency with earlier work, the predictions are generated using solar and lunar ephemerides from the Jet Propulsion Laboratory DE200 and LE200 ephemerides. Center of mass positions for the Sun and Moon are used with no corrections for center of figure, lunar limb profile or atmospheric refraction. The predictions depart from normal IAU convention through the use of a smaller constant for the mean lunar radius "k" for all umbral contacts. Times are expressed in either Terrestrial Dynamical Time (TDT) or in Universal Time (UT).

4. Internet Access and NASA Eclipse Bulletins

NASA eclipse bulletins can be read or downloaded via the World-Wide Web using a Web browser (e.g.: Mosaic, Netscape, Microsoft Explorer, etc.) from the GSFC SDAC (Solar Data Analysis Center) eclipse information page:

http://umbra.nascom.nasa.gov/eclipse/index.html

The original Microsoft Word text files and PICT figures (Macintosh format) are also available via anonymous ftp. They are stored as BinHex-encoded, StuffIt-compressed Mac folders with .hqx suffixes. For PC's, the text is available in a zip-compressed format in files with the .zip suffix. There are three sub directories for figures (GIF format), maps (JPEG format), and tables (html tables, easily readable as plain text). For example, NASA RP 1383 (Total Solar Eclipse of 1998 February 26 [=980226]) has a directory for these files is as follows:

file://umbra.nascom.nasa.gov/pub/eclipse/980226/RP1383text.hqx
file://umbra.nascom.nasa.gov/pub/eclipse/980226/RP1383PICTs.hqx
file://umbra.nascom.nasa.gov/pub/eclipse/980226/ec980226.zip
file://umbra.nascom.nasa.gov/pub/eclipse/980226/figures   (directory with GIF's)
file://umbra.nascom.nasa.gov/pub/eclipse/980226/maps   (directory with JPEG's)
file://umbra.nascom.nasa.gov/pub/eclipse/980226/tables   (directory with html's)

TABLE 1

Physical Ephemeris of the Umbral Shadow
Total Solar Eclipse of 1999 August 11

Universal Time	Center Line Latitude	Center Line Longitude	Diameter Ratio	Eclipse Obscur.	Sun Alt °	Sun Azm °	Path Width km	Major Axis km	Minor Axis km	Umbra Veloc. km/s	Central Durat.
09:30.3	41°02.0′N	065°05.4′W	1.0143	1.0287	0.0	69.5	60.8	–	49.0	–	00m46.5s
09:35	45°58.6′N	045°41.4′W	1.0189	1.0382	15.6	83.9	78.9	241.0	64.6	2.856	01m09.1s
09:40	47°41.6′N	037°00.3′W	1.0208	1.0421	22.5	91.6	85.8	186.4	71.1	1.922	01m20.4s
09:45	48°44.6′N	030°18.7′W	1.0222	1.0449	27.6	98.3	90.5	163.4	75.7	1.531	01m29.4s
09:50	49°25.8′N	024°37.3′W	1.0233	1.0472	32.0	104.4	94.1	150.2	79.4	1.305	01m37.0s
09:55	49°52.0′N	019°34.4′W	1.0243	1.0491	35.7	110.3	97.1	141.3	82.5	1.156	01m43.8s
10:00	50°07.0′N	014°59.3′W	1.0250	1.0507	39.1	116.1	99.5	135.0	85.0	1.049	01m49.9s
10:05	50°12.9′N	010°45.6′W	1.0257	1.0520	42.1	121.8	101.6	130.2	87.3	0.969	01m55.4s
10:10	50°11.3′N	006°49.4′W	1.0263	1.0532	44.8	127.6	103.3	126.4	89.1	0.906	02m00.3s
10:15	50°03.1′N	003°08.0′W	1.0268	1.0542	47.3	133.4	104.9	123.4	90.8	0.857	02m04.7s
10:20	49°49.4′N	000°20.6′E	1.0272	1.0551	49.6	139.3	106.2	121.0	92.2	0.816	02m08.6s
10:25	49°30.6′N	003°37.9′E	1.0275	1.0558	51.6	145.4	107.4	119.0	93.4	0.784	02m12.1s
10:30	49°07.3′N	006°45.3′E	1.0278	1.0565	53.4	151.6	108.4	117.4	94.4	0.757	02m15.0s
10:35	48°39.9′N	009°43.7′E	1.0281	1.0570	55.0	158.0	109.3	116.1	95.2	0.735	02m17.5s
10:40	48°08.9′N	012°34.1′E	1.0283	1.0574	56.4	164.6	110.1	115.0	95.8	0.718	02m19.5s
10:45	47°34.4′N	015°17.2′E	1.0284	1.0577	57.5	171.4	110.7	114.1	96.3	0.704	02m21.1s
10:50	46°56.7′N	017°53.9′E	1.0285	1.0579	58.4	178.3	111.3	113.5	96.6	0.693	02m22.2s
10:55	46°16.1′N	020°24.8′E	1.0286	1.0580	59.0	185.3	111.7	113.0	96.8	0.686	02m22.8s
11:00	45°32.6′N	022°50.6′E	1.0286	1.0580	59.3	192.4	112.1	112.7	96.9	0.681	02m23.0s
11:05	44°46.5′N	025°12.0′E	1.0286	1.0580	59.3	199.4	112.3	112.5	96.8	0.679	02m22.7s
11:10	43°57.8′N	027°29.6′E	1.0285	1.0578	59.1	206.4	112.5	112.5	96.5	0.680	02m22.0s
11:15	43°06.5′N	029°44.0′E	1.0284	1.0576	58.6	213.1	112.5	112.6	96.1	0.683	02m20.9s
11:20	42°12.9′N	031°55.9′E	1.0282	1.0572	57.8	219.6	112.5	113.0	95.6	0.690	02m19.3s
11:25	41°16.7′N	034°06.0′E	1.0280	1.0568	56.7	225.8	112.2	113.5	94.9	0.699	02m17.3s
11:30	40°18.1′N	036°15.1′E	1.0278	1.0563	55.5	231.6	111.9	114.2	94.1	0.712	02m14.9s
11:35	39°17.0′N	038°23.8′E	1.0274	1.0556	53.9	237.1	111.4	115.1	93.0	0.728	02m12.0s
11:40	38°13.2′N	040°33.2′E	1.0271	1.0549	52.2	242.2	110.7	116.3	91.8	0.750	02m08.8s
11:45	37°06.6′N	042°44.1′E	1.0267	1.0540	50.2	247.0	109.8	117.8	90.5	0.776	02m05.1s
11:50	35°57.0′N	044°57.8′E	1.0262	1.0531	48.0	251.5	108.6	119.7	88.9	0.810	02m01.0s
11:55	34°44.0′N	047°15.7′E	1.0256	1.0519	45.5	255.7	107.2	122.0	87.1	0.852	01m56.5s
12:00	33°27.3′N	049°39.6′E	1.0250	1.0507	42.9	259.5	105.3	125.0	85.0	0.905	01m51.5s
12:05	32°06.2′N	052°11.9′E	1.0243	1.0492	39.9	263.2	103.1	128.9	82.7	0.974	01m46.0s
12:10	30°39.9′N	054°55.9′E	1.0235	1.0475	36.7	266.6	100.3	134.0	79.9	1.067	01m40.0s
12:15	29°06.9′N	057°56.5′E	1.0225	1.0456	33.0	269.9	96.9	141.0	76.8	1.196	01m33.4s
12:20	27°24.9′N	061°21.8′E	1.0214	1.0433	28.9	273.1	92.5	151.5	73.1	1.390	01m26.1s
12:25	25°29.7′N	065°27.0′E	1.0200	1.0405	23.9	276.2	86.9	168.9	68.5	1.714	01m17.7s
12:30	23°10.9′N	070°48.3′E	1.0182	1.0368	17.7	279.5	79.0	206.0	62.4	2.412	01m07.4s
12:35	19°38.5′N	080°24.0′E	1.0151	1.0304	7.1	283.8	64.8	422.7	51.7	6.535	00m51.6s
12:35.9	17°33.5′N	087°17.2′E	1.0130	1.0261	0.0	286.1	55.2	–	44.6	–	00m42.3s

203

Total Solar Eclipse of 1999 August 11

FIGURE 1: THE ECLIPSE PATH THROUGH EUROPE

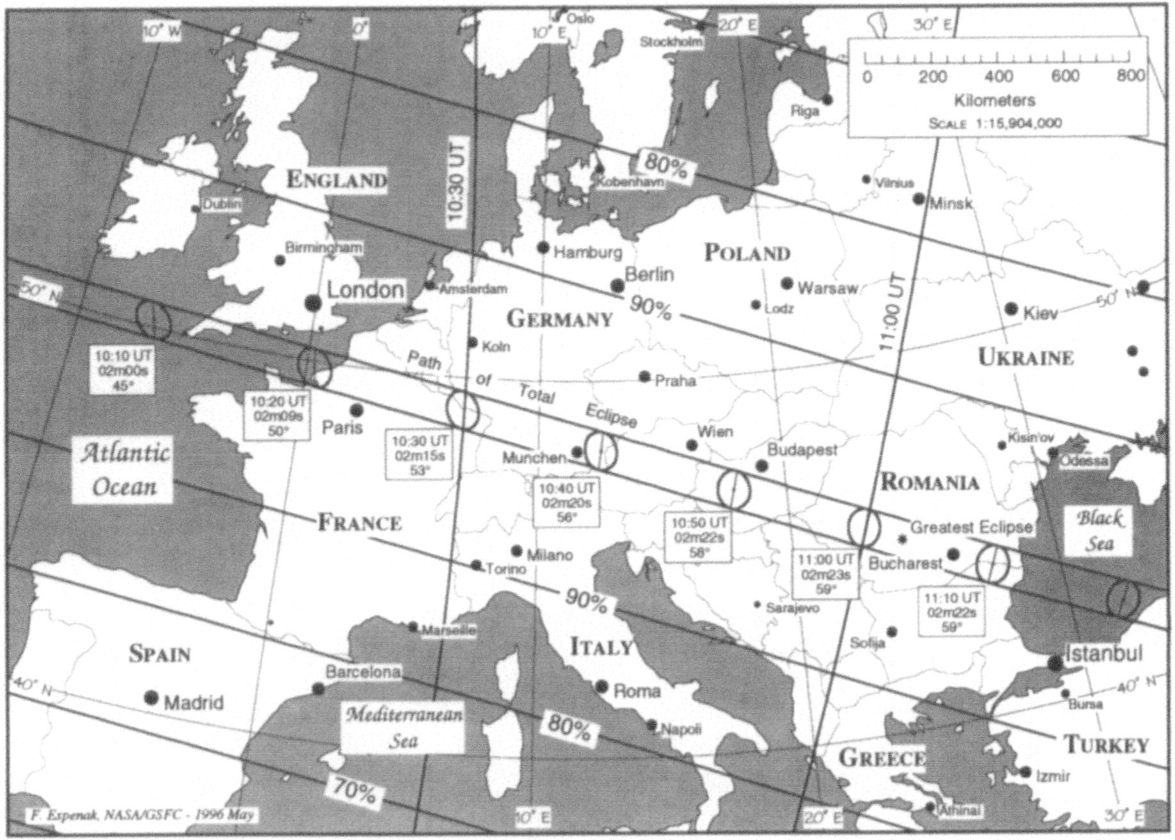

Figure 1

Total Solar Eclipse of 1999 August 11

FIGURE 2: THE ECLIPSE PATH THROUGH THE MIDDLE EAST

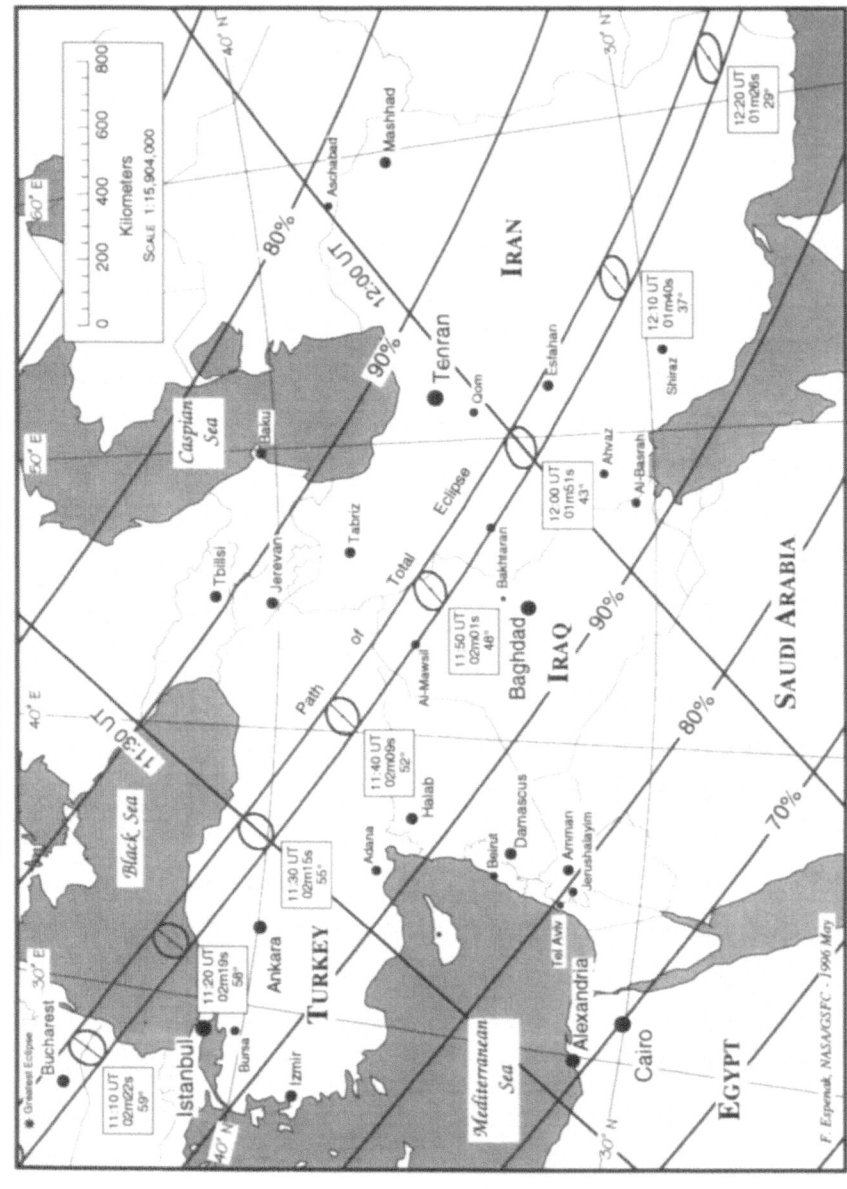

Figure 2

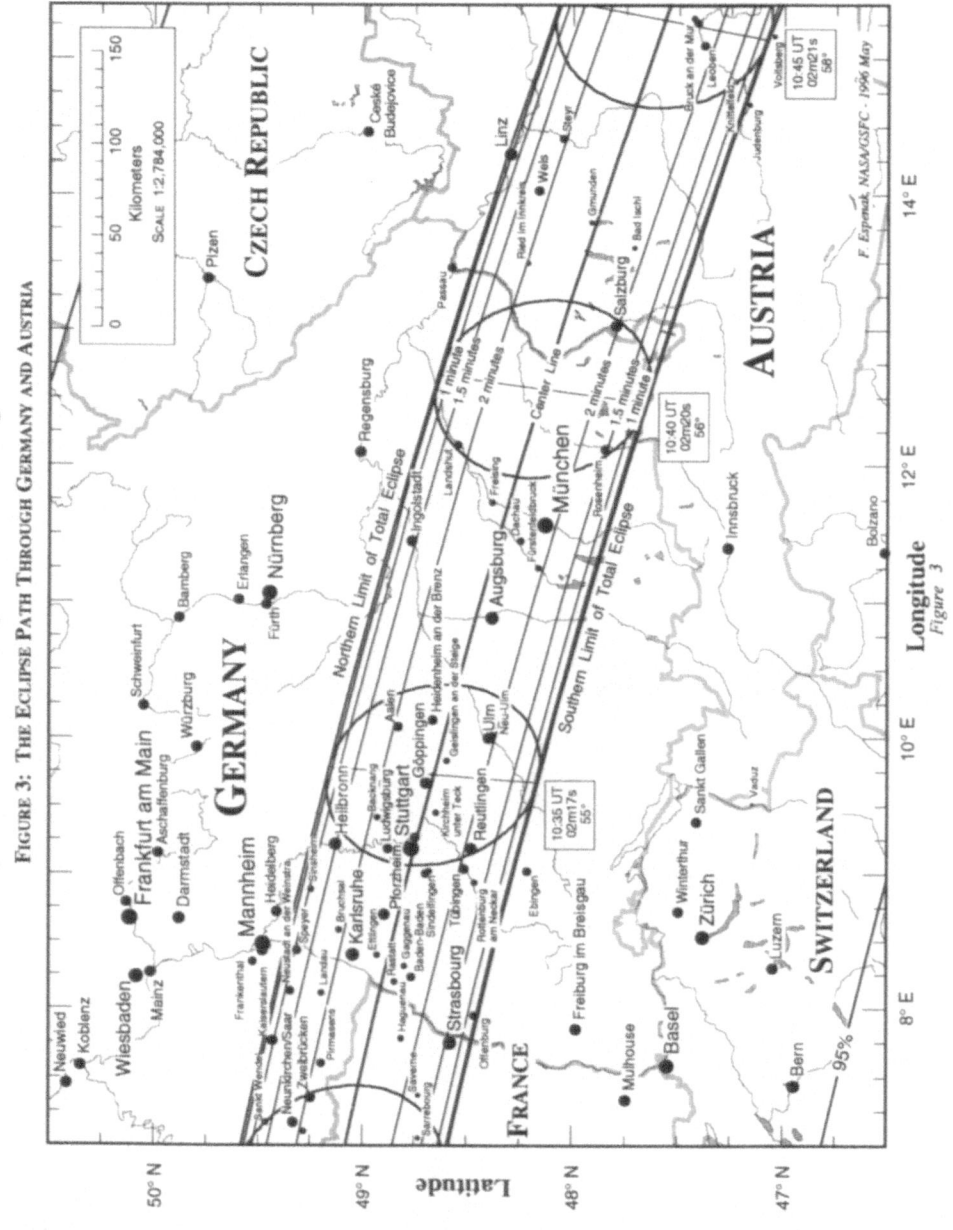

Total Solar Eclipse of 1999 August 11

FIGURE 3: THE ECLIPSE PATH THROUGH GERMANY AND AUSTRIA

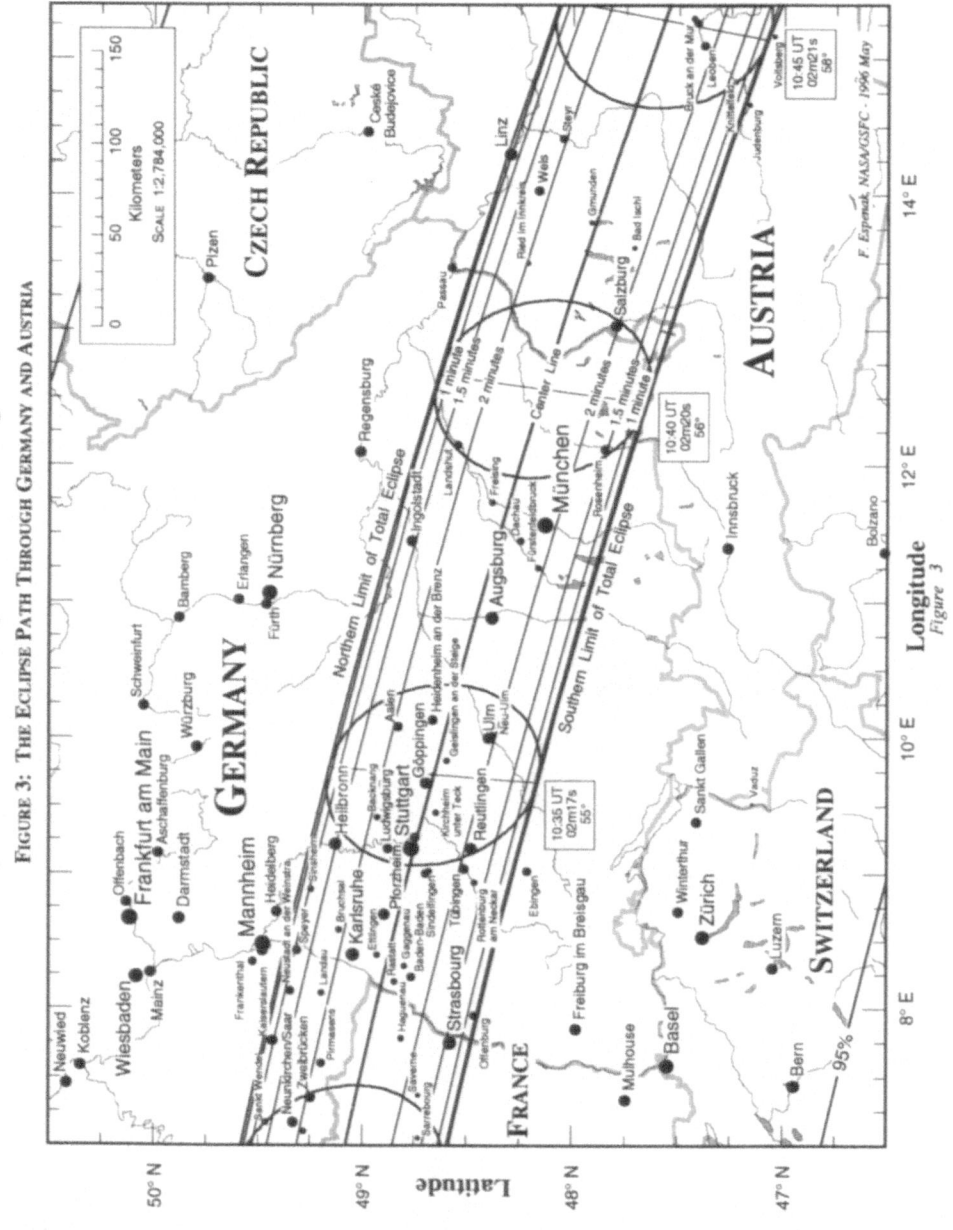

205

Total Solar Eclipse of 1999 August 11

FIGURE 4: THE ECLIPSE PATH THROUGH ROMANIA AND BULGARIA

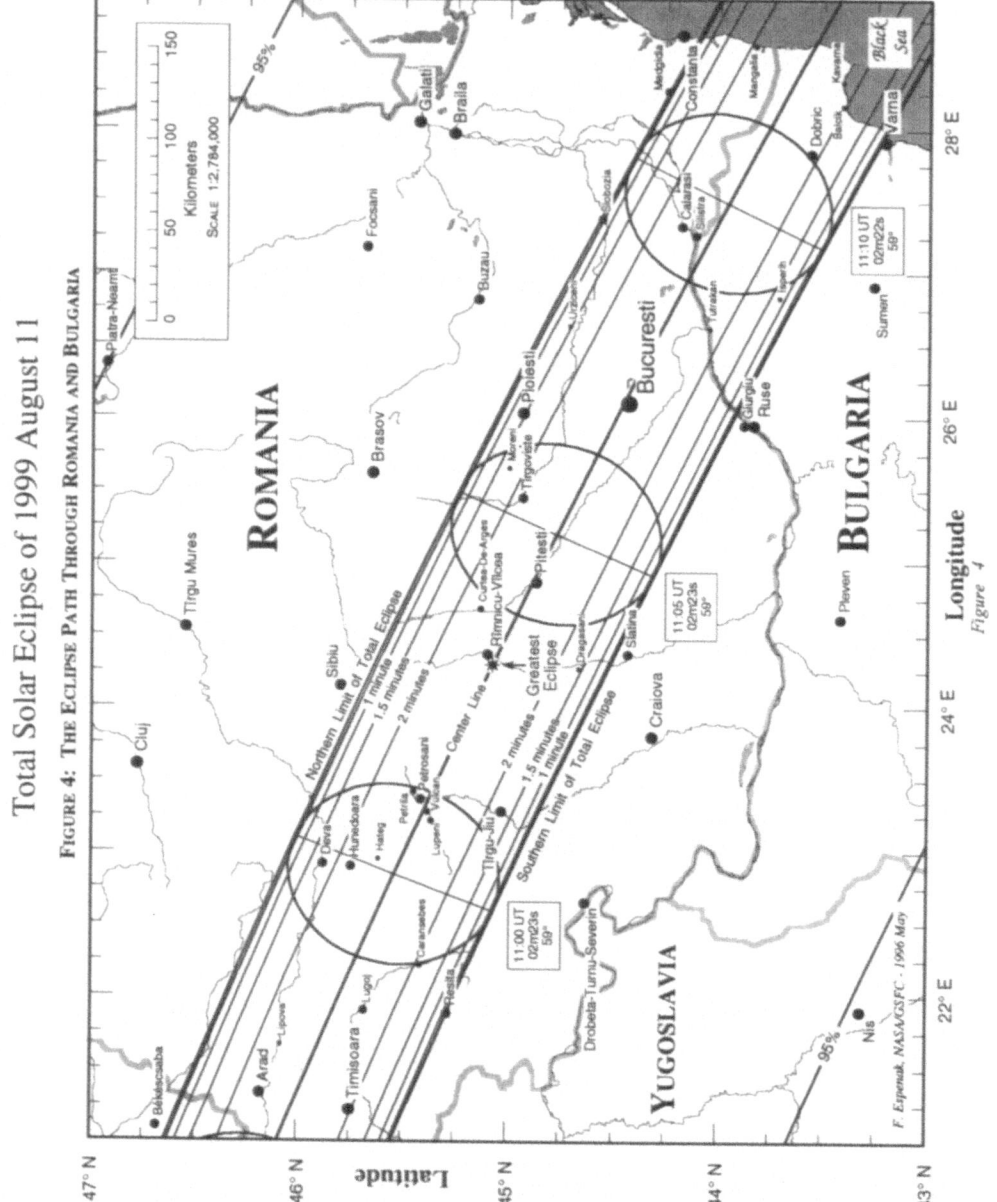

Files containing paths for every central solar eclipse from 1995 through 2005 are also available via the Internet. Each path is computed at 2 minute intervals of time. The path files can be accessed through GSFC/SDAC eclipse page, or directly at URL: http://umbra.nascom.nasa.gov/eclipse/predictions/yr-mth-day.html where "yr-mth-day" is replaced with the desired date (e.g.: "1998-february-26").

5. Ordering NASA's 1999 Eclipse Bulletin

NASA's bulletin for the total solar eclipse of 1999 will be available by the end of 1996. All participants of the NATO ARW on solar eclipses will receive a copy at that time. Others may order a copy by writing to the author or by sending a request via e-mail to espenak@lepvax.gsfc.nasa.gov.

References

1. Espenak, F. and Anderson, J. (1996) Predictions for the Total Solar Eclipse of 1999 August 11, NASA RP (in preparation), Washington DC.

ROMANIANS PREPARATIONS FOR THE '99 ECLIPSE

M. STAVINSCHI

Astronomical Institute of the Romanian Academy
5, Str. Cutitul de Argint, 75212, Bucharest, Romania

1. Present Situation of Astronomy in Romania

Six years ago, Romania experienced a great political and social upheaval that drew the whole world's attention. From the very beginning, astronomers tried to effect changes that would turn around the hopeless situation of their science. They hoped that the Great Change would bring the improvement so much expected in the life of the astronomical community. Less than two weeks after the Revolution, the leadership of the Institute was replaced and researchers began to search for the best way to emerge from the eclipse of Romanian astronomythat had lasted almost half a century.

The very few, astronomers (only 45 today) that remained collected the three Observatories in which they worked - in Bucharest, Cluj and Timisoara - together into a single Institute under the care of the Romanian Academy.

Isolated from the rest of the scientific world (both Western and Eastern) for decades, Romanian astronomers realized that their first priority was to learn about what their colleagues "over there", were done and form *a new generation of specialists*. So, twelve young people from the Institute left for the greatest academic centers in the world (in Belgium, France, Germany, USA) to acquire a specialization and, at the same time, a new conception about Science and society.

The next step: *computers*. Starting with none in 1990, we managed to construct a satisfactory computational base, and now even own a UNIX-type system based on a Power Challenge M (SGI) supercomputer with large central storage (RAM > 64 Mb) and an INDY PC100 workstation.

The connection to INTERNET (still unsatisfactory) and trips abroad have helped heal an old and extremely sore wound - *documentation*. A significant contribution in this context was the publication of our own scientific periodical - the *Romanian Astronomical Journal*.

Obviously, the hardest problem to solve was the *instruments*. All were old and located practically in the town. Some were shut down for good and consigned to museums. Some were closed only temporarily (the great meridian circle), for modernization to make them competitive with similar instruments. Some other instruments (the great astrograph or the 50 cm Cassegrain telescope) were equipped with CCD receivers, allowing successful participation in far-reaching international campaigns. Some instruments were moved in

Z. Mouradian and M. Stavinschi (eds.),
Theoretical and Observational Problems Related to Solar Eclipses, 209-213.
©1997 *Kluwer Academic Publishers.*

from higher latitudes and from light-bathed sites (like the Dunjon astrolabe from Brussels) to be fully modernized in Romania.

In spite of our efforts, getting money for astronomy in a country struggling in transition from one form of economy to another is impossible, and even a little ridiculous. Only a miracle could save us.

2. Total Solar Eclipse of 1999

Well, this miracle is the total Solar Eclipse of 1999. Standing right in the middle of such a rare event standing astride two millenaia (not to say the end of the world), with the certainty of no repetition for more than two centuries, is a genuine miracle.

The *maximum* of this eclipse will be in Romania. The Sun's maximum coverage (103%) will be in Romania and maximum eclipse duration 2 min 23 sec.

Unfortunately, our experience in this domain is zero. In 1961, when another total solar eclipse was visible from Romania, most of the today's "old people" of the Institute were students. To leave for another site on the globe - not in the least! (Before, we could not get permission to cross theborder. Now we have permission, but no money).

3. Past Experience

The situation was no easier in the past. There were some Romanian astronomers who had the chance to participate in such observations: N.Coculescu - the first director of the Bucharest Observatory - was sent in 1893 by the Romanian Government to accompany the French team in Senegal. His results are mentioned in the *Rapport annuel de l'Observatoire de Paris - 1893*.

There is not enough room here to pursue a real Romanian recordman in observing eclipses, N. Donici. At the age of 26, he observed the total solar eclipse at Elche, Spain. One yearlater, he left for Padang to observe the eclipse of 17/18 May 1901. Three years later he chose Pnom-Penh as observation site. The eclipse of 29/30 August 1905 found him in Spain again, at Alcala de Chisvert. On 31 August 1932, he observed an eclipse at Cap Porpoise; he was there for the 4th Congress of the International Astronomical Union (Cambridge, Massachusetts). We must mention that all results of his observations were reported systematically to the competent commissions, first with the International Union for Solar Studies and then with the IAU.

There were his other attempts to observe eclipses, but his singular fate moved N. Donici from Bessarabia (where he was born) to Russia (until 1917), then to Odessa (until 1920), to Romanian Bessarabia again (until 1940), to Romania (until 1944) and, lastly, to France, where his traces are lost.

The renowned astrometrist G.Demetrescu, however, used the opportunity of his presence in Paris to observe, along with Croze, the eclipse of 17 April 1912.

The last eclipse observed by Romanians was that of 1961, though unfortunately under bad weather conditions, by the Solar Department founded by Calin Popovici.

I omitted - purposely, of course - any contribution in solar eclipse research due to another specialist, that we persist in considering him being of ours, Zadig Mouradian.

Well, on the basis of the experience of the NATO AR Workshop participants in June at Sinaia, with our good will and enthusiasm, let us see what we have to do in order to be ready when the day of 11 August 1999 comes.

4. Band of Totality in Romania.

Since the short time spent by the workshop participants in Romania did not allow them to choose the best observation site,- I will try to introduce the "totality band" briefly here.

This band crosses variousareas,which are all very different inrelief, and all very attractive from a tourist standpoint. (Unfortunately,certain famous tourist attractions, like the Northern Moldavian monasteries and "Dracula's Castle", are outside the totality band.)

The eclipse enters the country in the Banat region, whose main city is Timisoara. Located at only 30-40 km from the slopes of Banat Mountains, which can be seen towards the East and South-East, the town was mentioned in documents as early as 1212 AD. Now it is well known as a cultural, commercial, industrial ,historical, and political center.

Fifty-three kilometers to the north, on Mures riverside, there is an even older town Arad.

A special attraction will be the mountains crossed by the eclipse: Poiana Rusca, Lotru Mountains, but first of all the Retezat Mountains, where the Romanian Academy has a natural reservation. The highest peaks are Peleaga (2509 m) and Parang (2518 m). Up there we could be nearest to the Sun during the early 1400's, and Targu Jiu - the place where the genius of the immortal sculptor Constantin Brancusi raised his famous column towards the skies.

Anyone who cannot climb the mountains, or wants to avoid the crowds of the cities, will perhaps watch the event from the Subcarpathian Hills, strewn with beautiful monasteries, especially along the Olt river.

Although less picturesque, the fertile Baragan Plain offers many facilities for observation camps. As regards our country's Capital population of about 2.5 million, even those which are not prepared for a genuine astronomical observation can watch the event from the balcony of their hotel, while the specialists will benefit by the, two solar refractors the only instruments especially meant for observing the Sun.

Nevertheless, while the mountains offer the advantages of heights and marvelous landscape, the hills their picturesque, genuine Romanian villages and with the quiet of the monasteries, one competitor hard to defeat seems to be the Black Seacoast with its very stable weather in August, a famous cliff, general comfort, and - last but not least - the rare opportunity of "basking in the sun" during the short "night" of the eclipse.

There is nothing else to do but come to Romania from now until 1999, to decide wich places would be best to install instruments and devices.

5. What Do We Have to Do Before 1999?

This is an extremely difficult question, considering the three factors we have already mentioned:

TABELE 1. Local conditions of the eclipse in different cities of Romania

n°	lat. ° '	long. ° '	Name	Duration of the totality phase m s	Maximum of the eclipse			
					UT h m s	g	h °	a °
1	+46 10	-21 19	ARAD	2 20.3	10 56 42.5	1.010	59	8
2	+44 25	-26 7	BUCHAREST	2 26.5	10 7 2.5	1.014	59	22
3	+44 51	-24 51	PITESTI	2 26.8	11 4 18.4	1.014	59	18
4	+44 57	-26 1	PLOIESTI	1 27.6	11 6 15.2	1.003	59	21
5	+45 16	-21 55	RESITA	0 28.1	10 58 36.5	1.000	60	10
6	+45 45	-21 15	TIMISOARA	2 9.6	10 56 57.6	1.008	59	8

n	First contact			Second contact			Third contact			Fourth contact		
	UT h m s	P °	Z °	UT h m s	P °	Z °	UT h m s	P °	Z °	UT h m s	P °	Z °
1	9 32 8.3	287	308	10 55 32.3	89	83	10 57 52.6	234	228	12 19 14.5	110	81
2	9 41 28.0	288	303	11 5 49.2	67	51	11 8 15.7	254	237	12 28 30.9	110	74
3	9 38 54.4	288	305	11 3 4.9	67	54	11 5 31.7	254	240	12 26 20.4	110	75
4	9 41 2.8	288	302	11 5 31.4	124	108	11 6 59.0	197	181	12 27 46.9	111	75
5	9 33 22.8	288	310	10 58 22.3	352	345	10 58 50.5	330	323	12 21 27.4	109	78
6	9 32 52.8	288	310	10 55 52.7	43	37	10 58 2.3	280	274	12 19 45.1	109	80

- precarious financial situation;
- small number of researchers;
- lack of experience.

There is nonetheless another factor, that cannot be neglected: enthusiasm.

So, our plan is ambitious and involves several aspects, not just scientific,but also researchers training by:

participation inthe eclipse of March 1997 (Siberia) and, perhaps, in that of February 1998 (South America);

astroclimatic investigations along the totality band, to establish the best sites for instruments (action already started in 1995);

observation of the partial eclipse of October 1996;

training of students by special lectures in Solar Physics and by an Advanced Study Institute on this topic;

establishment of the main scientific programs for Romanian specialists in 1999, in accordance with human and technical possibilities;

- precise determination of the eclipse contacts for the whole totality band;
- study of foreign references, and publication of our own works in Romania;
- suitable technical endowment of the Bucharest and Timisoara Observatories.

To coclude with the scientific aspects, we may mention the organization of this NATO Advanced Research Workshop;

In ligth of the fact that such an event will be very impressive for the public, have also we thought about a scientific education program.

The first objective of this program is the completion of a planetarium with a 15-meter dome in Bucharest (actually, right in the Observatory park), for simulations of astronomical phenomena and conferences for 150 spectators. We are now developing the architectural and technical project. We hope that the Planetarium will be inaugurated in 1998.

Teachers for all educational levels will be trained, with slides, booklets, and other materials to explain the phenomenon (especially its "coincidence" with the ... "end of the world").

At the same time, students, schoolboys, and astronomy lovers in general will be concerned with astroclimatic studies, as well as with the observation of the eclipse.

Since the eclipse has much greater implications than those indicated here (just think of the enormous flux of tourists we expect), we have set up an Inter-ministerial Commission dealing with eclipse questions.

Finally, to cover the extremely large range of the problems which will arise, and collect the necessary funds for the scientific campaign, we have initiated the International Association "Eclipse '99".

We are hoping that this difficult program harbors no unforseen hindrances.

JOSO WORKING GROUP 7
A focal point for collaboration around the 1999 solar eclipse

F. CLETTE
Observatoire Royal de Belgique,
Brussels, Belgium

Abstract A new working group dedicated to the preparation of the August 11, 1999 total solar eclipse was created recently as part of the JOSO (Joint Organisation for Solar Observations). This European group was created in order to prompt scientific collaboration around this event. The aims and future actions of JOSO WG7 are outlined here. All research groups who are planning to set up a scientific programme at this occasion are invited to join this new community.

1. Introduction

During the last JOSO board meeting in October 1995, the idea of a new working group dedicated to solar eclipses was adopted. Indeed, the last total eclipse of the century is due to happen on August 11, 1999 across the densely populated European continent. As the JOSO acts as the consortium of solar observers from all European countries, it is best placed to promote collaboration and scientific coordination at the European level and with teams from other parts of the world.

The present composition of this working group (WG7) is the following : F. Clette (Belgium, Chair), V. Dermendjiev (Bulgaria), I.S. Kim (Russia), S. Koutchmy (France), V. Makarov (Russia), G. Maris (Rumania), Z. Mouradian (France), V. Ruvsin (Slovakia), I. Vince (Yugoslavia), A. Okten (Turkey). This list is not limited, and new members are invited to join in. If for instance, we just consider the countries crossed by the Moon's shadow, representatives from United Kingdom, Luxembourg, Germany, Austria and Hungary are still missing, at the time of this writing. Moreover, eclipse observers from other parts of Europe are also welcome.

2. Aims

As a starting point, some fundamental aims were defined for this working group:
1. WG7 should gather information concerning:
   - research groups wishing to organize an eclipse expedition somewhere in Europe;
   - local institutes, in countries where this eclipse takes place, that are willing to help visiting observers to carry out their scientific program at selected sites;

Z. Mouradian and M. Stavinschi (eds.),
Theoretical and Observational Problems Related to Solar Eclipses, 215-217.
©1997 *Kluwer Academic Publishers.*

•local circumstances of the eclipse, with an emphasis on recent climatological data.

2. WG7 must also promote collaborations, by bringing together teams with common interests, for instance by matching special requirements with resources and support made available at specific locations. This can be done most efficiently during meetings, like the present workshop or the annual JOSO meetings, and also through the Internet.

3. This working group should publicize the resources available in Europe for scientists from the rest of the world. In this respect, JOSO's WG7 might serve as the European node for the preparation of observing missions. From a more general point of view, this eclipse also offers an excellent occasion to make some ignored resources more widely known to the community of solar physicists, and to take some scientists out of their relative insulation, in particular in countries of Eastern Europe.

4. Later on, WG7 could also help in the organisation of a post-eclipse conference dedicated to the presentation of the 1999 eclipse results.

This tentative list of objectives is rather ambitious, and as such, it is open to modifications. For instance, following recent suggestions, two important aspects might be added and maybe take precedence over some of the above aims :

•the working group, backed up by the JOSO board, could facilitate the exportation of scientific equipment to the observing sites, through official contacts with local administrations and authorities;

•as the working group focuses specifically on that spectacular eclipse event, it might take part in the dissemination of basic but accurate information about solar eclipses and eclipse viewing to the general public, by various means : news media, Internet, or more officially, through the Departments of Education. This contribution to public education is fundamental, but poses a challenge in the context of Europe due to the multiplicity of languages. The translation by local astronomers of a reference pamphlet, edited in english, for their own country could be a solution. This exciting project would thus require the participation of many volunteers.

3.First actions

As only a few scientific teams have actually started to build up firm plans for the 1999 solar eclipse, the WG7 activities during its first months of existence will be concentrated around a few priorities, in order to provide first a foundation for the completion of the overall project.

A general call for participation will be issued, first to the JOSO community, and later to other scientific institutes, to obtain as wide a coverage as possible of the interest for this eclipse. This will lead to a first survey of the scientific teams and institutions involved in this future event, which in turn will form the basis for an "Eclipse 1999" database. In order to make this information about all aspects of this solar eclipse widely accessible, especially for the public and for scientists outside

Europe, a WG7 site will be created on the World Wide Web (with, possibly, an associated FTP server). This site could be implemented on the JOSO server, as part of the existing JOSO site, with the help of Dr. M. Messerotti.

All these activities will take place in coordination with the Eclipse Section of the International Astronomical Union (Chairman : Dr. J.M.~Pasachoff), in order to avoid any duplication and overlap. In this respect, WG7 would serve in part as a local branch of this world-wide institution at the occasion of this once-in-a-lifetime European total eclipse.

Contact address :
Dr. F. Clette \\
Chairman JOSO WG7
Observatoire Royal de Belgique
Avenue Circulaire, 3
B-1180 Bruxelles
Belgium
E-mail : fred@oma.be

A RADIO OBSERVATION PROJECT FOR THE 1999 SOLAR ECLIPSE

A. ONCICA
Astronomical Institute of Romanian Academy
Str. Cutitul de Argint 5, 75212 Bucharest - Romania

1. Introduction

The spatial and spectral features of the solar atmosphere make a certain class of observation easier during a solar eclipsethan unterb other conditions. When the transits of the East and West moon limbs are viewed as moving knife-edge diffraction screens the corresponding diffraction pattern is an expression in the time domain of the 2D spatial information of an equivalent interferometer, with all baselines sampled simultaneously.

The angular resolutionthat can be obtained in the centimeter-to-meter range is roughly one to ten arcseconds at the minimum, which corresponds to solar structures in the 500--5000 km range. For a ground-based radio interferometer to attain the same angular resolution would require baselines of 3--30 km or more (hardly to be found by chance on or near the path of totality). With multi-frequency observation, it is possible to attain, at least in principle the third dimension through the height above the photosphere dependence of local electron density (i.e. plasma frequency).

2. Lunar Occultation

Occultation of stars by the Moon has long been used as a high-resolution method of studying stellar properties. The first use for radio sources came at the very start of modern radio astronomy, when Dicke and Beringer observed the partial solar eclipse of 9 July 1945 at Cambridge, Mass. at 1.25 cm wavelength. Eclipse observations have since been made to study both the quiet sun and the radiation associated with solar active regions.

For the circumstances of an occultation the Moon's limb is assumed to be smooth and to have a very large radius of curvature. The diffraction pattern on the ground is produced by interference between beams from all elementary sections of the wavefront. The situation is analogous to an interferometer with all baselines sampled simultaneously.

Let $D=3.84 \times 10^8$ m be the (mean) distance to the Moon and λ_0 the frequency of observation. The important parameter is θ_0, the angular size of the first Fresnel

$$\text{zone:} \, \theta_0 = \sqrt{\frac{\lambda_0}{2D}} \qquad (1)$$

219

Z. Mouradian and M. Stavinschi (eds.),
Theoretical and Observational Problems Related to Solar Eclipses, 219-222.
©1997 *Kluwer Academic Publishers.*

As an example, at $\lambda=10$ cm, $\theta_0 =1.14\times10^{-5}$ rad$=2,35$ arcsec and scales as $\lambda^{1/2}$. This resolution is equivalent with a terrestrial *filled* baseline of 9 km. The knife-edge diffraction pattern of a point source (in units of $x= \theta/\theta_0$) can be compared with the classical geometrical shadow of a Heaviside shape (figure 1).

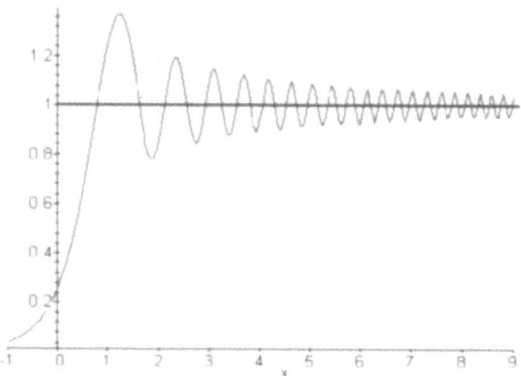

Figure 1: Knife-edge diffraction pattern of a point source

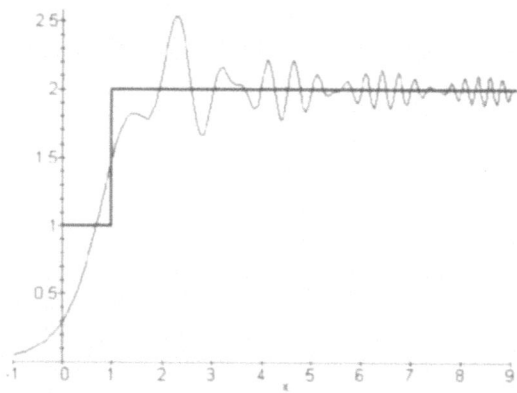

Figure 2: Normalized diffraction pattern for two identical point sources separated by an angular distance equal to the first Fresnel region

Note that the amplitude of the first few fringes are of the order of 10% of the geometrical amplitude. An extended radio source will produce a smeared diffraction pattern. For two identical point sources separated by θ_0, the diffraction pattern has the distinct shape of figure 2.

If we allow for multi-frequency observations, then for every λ, we will have a diffraction pattern whose fringe visibility will depend on λ_0. Keeping the wavelength of the previous diffraction pattern as unit, we build a diffraction surface depending on a new parameter: $y=\lambda/\lambda_0$ (see figure 3).

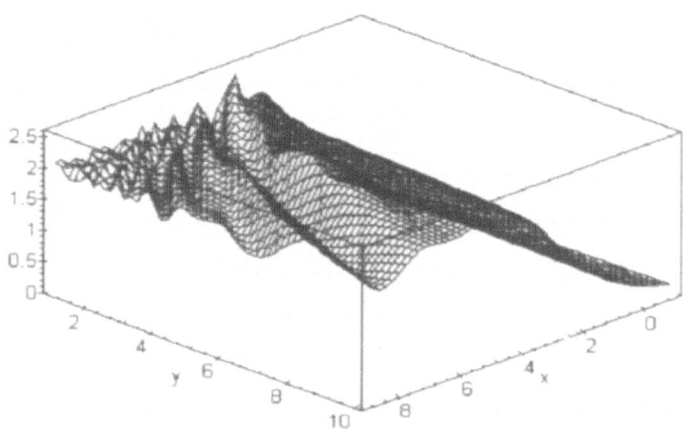

Figure 3: Two dimensional (angular and spectral) diffraction pattern

It can easily be seen that, at short wavelengths (y=1), the two sources are resolved; but at longer wavelengths, the diffraction pattern is smeared out.

To see what is the relation of figure 3 with the recorded solar dynamic spectrograph we must remember that the Moon's limb sweepe across the Sun at a rate of approximately 0,5 arcseconds/s. Returning to the numerical example given above, x is in units of 4.7s of time and λ ranges from $\lambda_0 =10$ cm to $10 \times \lambda_0 =100$ cm.

3. Resolution Limits

To evaluate the effects of and frequency resolution on spatial resolution, we shall follow a procedure described in [1] and [2], but with a somewhat different goal. We shall look for the necessary conditions to resolve the first k fringes up to $\varphi=\theta_0/k$ fringe spacing

The receiver time constant must be smaller than the smearing time for the Fourier component at $1/\psi$ cycles per radian:

$$\tau \leq \varphi / \overset{\bullet}{\theta} \cong 2 / k \quad \text{sec} \tag{2}$$

For the given example, this is evaluated at (4.7/N) s. If the time constant proves to be too small, it one can always be increased it by smoothing the data later.

The receiver frequency bandwidth B is limited by the smearing of the highest desired Fourier component. For small relative bandwidths:

$$B \leq 3f^2 \varphi^2 D / c \cong f / k^2 \tag{3}$$

For single frequency observations the frequency bandwidth is generally fixed. If it is too narrow, you lose sensitivity. If it is too wide, you lose spatial resolution. But with a dynamic spectrograph there is no such problem. You can work at the highest spectral resolution and still smooth the *spectrum* later taking the mean of adjacent channels.

The signal-to-noise ratio will depend on whether or not the source is resolved. For the unresolved source, and for optimum time constant and frequency bandwidth, the signal-to noise ratio is :

$$\frac{S}{N} \cong \frac{T_A}{T_N} f\varphi^{3/2} \sqrt{\frac{3D}{2\pi c\,\theta}} \cong 2 \cdot 10^3 \frac{T_A}{T_N} \sqrt[4]{\frac{f_{(MHz)}}{k^3}} \tag{4}$$

Note that the signal-to-noise ratio increases with the observed frequency. This is the consequence of increased optimum bandwidth. The system noise temperature includes a substantial contribution from the Moon. Here we can see that for small aperture instruments the real limiting factor is the signal-to-noise ratio. But typical solar sources with fluxes of the order of few s.f.u can contribute a few kelvin to the antenna temperature well within the range of typical sensitivities.

4. Conclusions

A solar radio spectrometer can be used for observatuions with good spectral and time resolution. During a solar eclipse, the time and frequency resolutions can be used to probe the 2-dimensional angular structure of the sources. The spectral content of such a dynamic solar spectrum can also bring information on the third dimension, through the height above photosphere-dependence of the emitted radiation spectrum. As the path of the August 11, 1999 total solar eclipse comes near most European solar radio instruments of this kind this will provide a good opportunity for such observations. It is also worth mentioning here that radio observations are much less sensitive to atmospheric conditions than opticalobservations are.

References

1. Hazard C. (1976) Lunar Occultation Measurements, in Meeks M. L. (ed.) *Methods of Experimental Physics*, Vol. **12**, *Astrophysics*, Part C, *Radio Observations*, Academic Press, New York, 92-117.
2. Cohen, M. H. (1969) High-resolution Observation of Radio Sources, *Ann. Rev. of Astron. and Astrophys.* **7**, 619-664.

ROMANIAN PROJECTS FOR THE 1999 TOTAL SOLAR ECLIPSE

G. MARIS
Astronomical Institute of the Romanian Academy
Cutitul de Argint 5,
75212 Bucharest 28, ROMANIA
e-mail: gmaris@roimar.imar.ro

Abstract This paper briefly reviews the activity of the Solar Department of the Astronomical Observatory in Bucharest and the instruments that are now available at the Observatories in Bucharest, Timisoara and Cluj. The observation projects for the total solar eclipse of 1999, based on the present instrumental possibilities, are presented.

1. Introduction

The totality band of the last total solar eclipse (TSE) of the 20th Century will pass through Romania, from West to South-East, in a strip 112 km wide. The maximum duration (2m 23sec) will be on Romanian teritory, as well as the maximum coverage (103%).

A total solar eclipse is an astronomical event that is seen very rarely from the same place on Earth. Even more unusual is when an Astronomical Observatory has the happy chance of being located on the totality bands of two TSEs in a time interval of no more than 39 years. This is the case of the Bucharest Observatory, which was 70 km North of the center line of February 15, 1961 TSE, and will be situated exactly on the center line of the August 11, 1999 eclipse. At this eclipse, the Timisoara Astronomical Observatory is also located in the band of totality, at about 25 km South of the center line (Figure 1).

In this very favourable situation for the two Observatories, we have to come up with some projects for the TSE '99 observations, with the stable instruments.

Before presenting our instruments I should sketch a short history of solar research in Romania, after which our scientific projects will be reviewed including both the ground-based and spaceexperiments.

2. Solar Research in Romania

Astronomical Institute of the Romanian Academy of Sciences is composed by three Astronomical Observatories: Bucharest, Cluj and Timisoara. Research and observational work in solar physics is performed only in the Bucharest Observatory. Our Solar Department began 41 years ago, in preparation for the International

Z. Mouradian and M. Stavinschi (eds.),
Theoretical and Observational Problems Related to Solar Eclipses, 223-228.
©1997 *Kluwer Academic Publishers.*

Figure 1. *The eclipse path through Romania.*

Geophysical Year (1957 - 1959), the first broad program of international cooperation for solar observations.

Two instruments were installed for this programme:

a) a Zeiss refractor (13/195 cm) with a photographic camera for white light observations of the solar photosphere;

b) a Zeiss refractor (8/120 cm) with a Lyot-Ohman H_α filter for daily photographic and visual patrol of active chromospheric phenomena.

Both these instruments are still operating.

We carried out the solar patrol observations of the photosphere and chromosphere from the very beginning. The data obtained on active chromospheric phenomena and on relative sunspot numbers and sunspot positions were reported monthly to the Solar World Data Centers (Boulder-Colorado, Meudon, Moscow, Zurich andlater,, Brussels) and to the Pulkovo Observatory. They are also published in our yearly bulletin, "Observations Solaires" edited by Romanian Academy Publishing House, since 1957.

Research was then developed in cooperation with other solar observatories, namely: Pulkovo Main Astronomical Observatory (Russia); Ondrejov Astronomical Observatory (Czech Republic); and, in recent years, with the Paris-Meudon Observatory and the Astronomical Institute, in Sofia (Bulgaria). We took part in certain international observational program: IQSY (International Quiet Sun Years); INTERKOSMOS Programmes; and KAPG (among former Eastern Europeean countries); CINOF,among others.

Our present staff consists of six researchers.

Over many years, we were concerned with the following topics in the solar and solar-terrestrial fields:

- statistical studies of solar activity phenomena during solar cycles;
- morphology and dynamics of the active regions;
- forecasts of the solar activity cycles;
- the statistical study of total solar irradiance variations and their correlations with certain solar activity indices;
- solar prominence spectroscopy using observational data from Ondrejov and Pulkovo Observatories;
- MHD modelling of the solar activity cycle, and of certain coronal phenomena (prominences and coronal stramers);
- coronal mass ejections: their associatons with other solar phenomena and, the geophysical effects of such complex of events;
- solar proton event distribution during the eleven-year solar cycle and their effects on Earth;
- coronal holes and high speed plasma streams in the solar wind;
- the study of the geomagnetic, ionospheric, and biologic effects of the solar activity phenomena.

More information about our activity during the first ten years can be found in [1] and, since 1991, in the JOSO Annual Reports [2, 3, 4, 5, 6].

As concerns our experience in total solar eclipse observations, we have practically none. Romania was covered by the totality area of the 15 February 1961

TSE. On that occasion, scientists had developed experiments, studied the observation conditions at different sites, and performed simulations. But the sky was clouded in all the selected observation places. All they managed to obtain were a few pictures of the solar corona, taken from a plane (IL 18) flying through the totality band, from West to East [7].

3. Instrumental Supply of the Romanian Observatories

Besides the two mentioned refractors, *solar department* of the *Bucharest Observatory* has also a number of other instruments which might be used for observations, such as:
- two Bardou refractors 10.8/165 cm;
- two Zeiss refractors 6.3/27 cm.

In the eighties, we built a horizontal spectrograph for students' use with:
- a 30 cm Jensch coelostat;
- a telescope objective lens of 20/225 cm;
- a Bausch and Lomb grating with 1200 grooves/mm;
- the collimator and the camera mirrors 10/200 cm.

The Bucharest Observatory also has the following instruments:
- a double astrograph of Prin-Mertz construction (38/600 cm) with a field of view of $2° \times 2°$; with a photographic camera that uses 24x24 cm plates and a 768 x 512 pixel CCD;
- a Cassegrain telescope (50/750cm), Zeiss-Jena make, with a photoelectric photometer and photomultiplier of type EMI 9502 and UBV-Johnson filters; using a 375 x 242 - pixel CCD.

The Cluj Observatory has a Newtonian telescope (50.8/250 cm) equipped with a photomultiplier, in an equatorial fork mounting with a refractor of Prin construction (20.32/300 cm).

The Timisoara Observatory has a Cassegrain telescope (30/169 cm) provided with a photoelectric photometer with a photomultiplier type EMI 9862 Q.

It is with these means that we are preparing the total solar eclipse of 11 August 1999.

4. Projects for the 1999 Total Solar Eclipse

In spite of the theoretical progress and the development of modern technologies, and inspite vof the fact that the Sun is continuously being watched by high-resolution ground-based telescopes and space missions that could not even be imagined 50 years ago (Yohkoh, Ulysses, SOHO), many questions remain concerning the solar corona such as:
- the coronal heating mechanisms to temperatures of 1-2 million K;
- the density and temperature distributions in coronal structures (coronal streamers, coronal holes, coronal plumes,and other);
- the nature of coronal rotation (differential or rigid);
- the origin of the solar wind.

We hope that many observing teams will come to us for their observational projects at the eclipse of 1999, and that these unsolved questions will be answered in the observations on Romania's territory. We ourselves might then have other possibilities for some sophisticated experiments.

Besides these hopes, based only on the present instruments, we have the following observational projects to prepare:

1. Fine structure in white light observations of the lower corona using radial filters, at the Bucharest and Timisoara Observatories.

2. Coronal structure observations at the distance of 1-5 solar radii using radial filters; we have in mind structures such as streamers, helmet, arches or even coronal mass ejections, considering that there will be maximum solar activity.

3. Coronal structure evolution. The corona will be observed at two different moments, using two Bardou refractors with the same optical characteristics, located one at Timisoara and the other at Bucharest; the time interval between two pictures will be of about 13 min. The two refractors could be situated at the extreme points of the center line (or nearby) on Romanian territory. Two photos could be used for studying of the structure of the inner corona.

4. Spicules observations.
4.1. Hα observations of spicules orientation at the second and the third contacts (Lippincot, 1957) [8].
4.2. Spicules spectra with a CCD camera in H and K lines of Ca II.

5. Monochromatic coronal observations in green line, red line and continuum, using interferential filters, with the Bucharest and Timisoara telescopes, in order to obtain an accurate determination of (dE/dr) by photometric processing

6. The determination of the contact times C1 and C4 (exterior contacts) at all the Observatories (Bucharest, Timisoara, Cluj), and, *the contact times C2 and C3* at the Bucharest and Timisoara Observatories.

7. Einstein effect. Using the Bucharest astrograph, we can take the images of the stellar field around the Sun during the total phase of the eclipse in order to determine the stellar positions accurately. The field of the instrument, the exposure time of the plate during the totality phase (about 0.5 sec), and the precision of the stellar position measurement by Ascorecord machine make such an experiment possible. Preliminary tests could be made beginning in August of this year.

8. IR observations. The Multichannel Infrared Spectrometer ISTOK-1 measuring the emission infrared spectrum in the 4-16 μm, using 64 separated channels, is working on the MIR space station. The optical system and detector array are placed on a two-axes oriented platform. Among the working regimes of the instrument are the platform orientation to the Sun and solar investigation in the medium infrared range.

We propose acquiring data in the mean infrared 4 - 16 μm during the eclipse. Such data can be used in analysis of the solar corona and of the circumsolar dust rings [9].

5. Conclusions

In 1996, three years before the big event, we are trying to outline some observational projects for which we shall use our existing instruments in the Observatories of the Astronomical Institute of the Romanian Academy.

Naturally, we shall not be able to obtain any significant results with our present instruments alone. Although the odds are not in our favour, we hope to succeed in acquiring CCDs for all the present instruments, as well as a new Hα filter and even a multiband filter.

The researchers of our Department are engaged in scientific projects of the SOHO mission, while their colleagues from the Gravity and Space Sciences Institute are co-authors to some IR experiments on MIR. That is why we hope that in the space research field we shall be able to initiate interesting programs of eclipse obsevations.

In this paper we have presented *only our projects, and would welcome any suggestions.*

We believe that the first of our projects, namely the bringing together of solar eclipse specialists at this meeting, has succeeded.

Thank you all for being here!

References

[1] Popovici, C. (1969), *Solar Phys.*,**9,** 494-495.

[2] Maris, G. (1992), *JOSO Annual Report*, 35-36.

[3] Maris, G. (1993), *JOSO Annual Report*, 42-43 .

[4] Maris, G., Oncica, A. (1994), *JOSO Annual Report*, 56-57.

[5] Maris, G. (1995), *JOSO Annual Report*, 39-40.

[6] Maris, G. (1996), *JOSO Annual Report*, (in press).

[7] Tifrea, E. (1996), in Proc. of NATO ARW *"Theoretical and Observational Problems Related to Solar Eclipses"*, Poster Papers, *Rom. Astron. J.*, (in press)

[8] Lippincot, S.L., (1957), *Smithsonian Contrib. Ap.*, **2**, no.2.

[9] Maris, O. (1996), in Proc. of NATO ARW *"Theoretical and Observational Problems Related to Solar Eclipse "*, Kluwer Academic Publ., (this issue)

EIT, THE EXTREME-ULTRAVIOLET IMAGING TELESCOPE ON BOARD SOHO: FIRST RESULTS.

P. CUGNON, F. CLETTE
Observatoire Royal de Belgique Avenue Circulaire, 3, B-1180 Bruxelles, Belgium

J.-P. DELABOUDINIERE
Institut d'Astrophysique Spatiale,91405 Orsay Cedex, France

EIT SCIENCE CONSORTIUM*

Abstract EIT is a EUV telescope on board SOHO providing wide-field images of the corona and of the transition region, on the solar disc and up to 0.5 R_0 above the limb. The multi-coated optics of the telescope select four emission lines to provide temperature diagnostics in a wide range, corresponding to the coronal plasma and the underlying cooler layer. This paper briefly presents the instrument and its first observations.

1. Description of the Instrument

EIT is a normal incidence telescope, with two mirrors in a Ritchey-Chr,tien configuration [1]. The four quadrants of the mirrors are coated with different EUV reflecting multilayers, providing narrow band images at four selected wavelengths with a spectral resolution between 10 and 100. A rotating mask in front of the telescope makes that only one sector is illuminated at a time.

The 1024 x 1024 pixels CCD is digitized to 14 bit, with a pixel size of 2.6 arcsec, and a total field of view 45 x 45 arcmin. Three successive aluminium filters prevent visible and UV light from reaching the CCD. The main telescope body consists in a cylindrical vacuum vessel with an air-tight front door. An external radiator passively cools the CCD down to - 70° C.

The four selected spectral bands are centred respectively on three coronal lines (Fe IX-X, 17.1 nm; Fe XII, 19.5 nm; Fe XV, 28.4 nm) and the He II at 30.4 nm produced in the upper-chromosphere and transition region. Each emission line marks a temperature interval within a temperature range from 4 x 104 K in the transition region to 3 x 106 K, a typical temperature for the hot plasma inside active regions. EIT is then perfectly adapted to the study of steady state processes which contribute to a significant part of the energy input and dissipation in the corona. Though not being properly speaking a flare instrument, EIT is nevertheless able to observe a wide range of energetic phenomena, a property which will be fully exploited in the ascent of cycle 23, and, hopefully, also during the maximum phase.

Z. Mouradian and M. Stavinschi (eds.),
Theoretical and Observational Problems Related to Solar Eclipses, 229-232.
©1997 *Kluwer Academic Publishers.*

2. Observational Program

The main observational modes are : full Sun high resolution (1024 x 1024) in all wavelengths, full Sun medium resolution (512 x 512) in all wavelengths and full resolution selected areas with high temporal resolution. The choice of the operating mode is highly dependant on the available telemetry. Presently, at least one set of full resolution images (Figure 1) is transmitted everyday. Fast-rate time series of up to 100 sub-images are also possible and have already been tested. In the future, the synoptic program may be somewhat extended in connection with the compression capabilities of the inboard software. The high temporal resolution series (up to a time resolution of 5 seconds in the best case) will take place in the context of event driven observations of active phenomena, or during coordinated campaigns implying other SOHO instruments and ground-based observatories [2].

3. Scientific Objectives

The primary scientific objective of EIT is to study the dynamics and evolution of coronal structures over a wide range of time scales and temperatures and to bring new insight in the mechanisms responsible for the coronal heating and the solar wind acceleration [3]. This includes the study of the long-term behaviour of global structures in the corona, the survey of the bright points distribution and of their connection with the underlying phenomena, the mapping and analysis of candidate sources of the solar wind, and the study of the evolution of the chromospheric network.

4. First Results

Thanks to its sensitivity to temperature, EIT is able to distinguish features at different plasma temperatures inside the same magnetic region of the corona. Simultaneous subfield images at 17.1 nm and 28.4 nm of an active region (Figure 2) show indeed well differentiated locations for their respective maxima of brightness. Time sequences of such images reveal fast local changes with time scale of a few seconds, and through the solar rotation will allow three-dimensional reconstructions. This, combined with the temperature diagnostic, will lead to an improved physical description of active regions.

The very good image contrast also allows unprecedented views of faint structures inside coronal holes and at their boundaries. The polar holes, particularily at the South pole, show in the three coronal iron lines a well developed network of polar plumes (Figure 1a & 1b). Most of these plumes seems connected to a bright point.

The chromospheric network is apparent on Figure 1d, and Figure 3 (He II 30.4 nm). Moreover, this last set of images shows the emergence of a huge arch-prominence with rapid evolution (the two intermediate images are separated by 25 min), with a likely re-organization of the magnetic configuration between April 30 and May 1. The polar holes visible at 30,4 nm also reveal a lot of activity with macro-spicules erupting as short-lived jets reaching altitudes up to 0.3 $R_\odot$ above the limb.

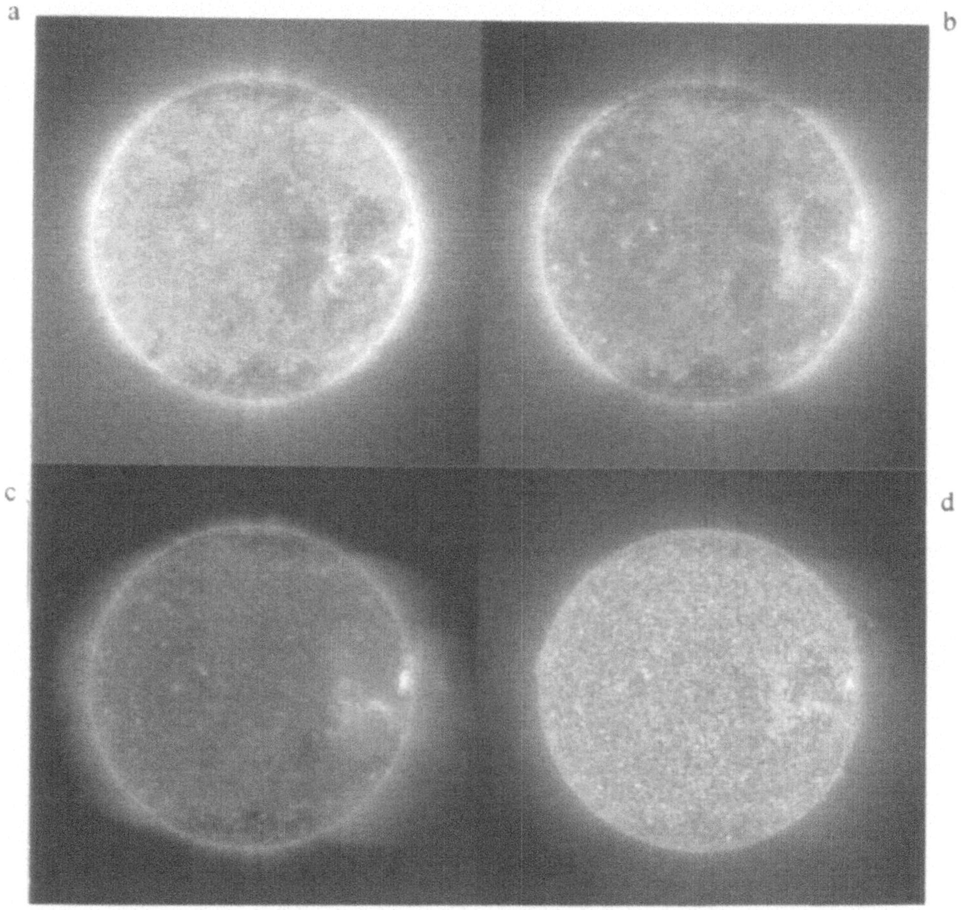

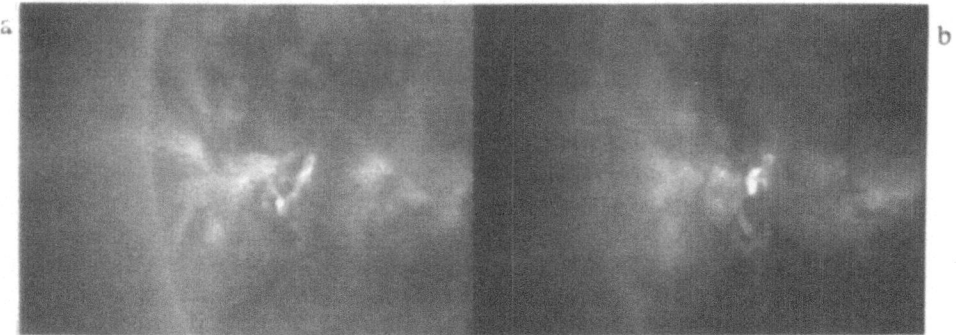

Figure 1 : Full-resolution images at the four selected wavelenghts, from top to bottom and left to right : (a) 17.1 nm, (b) 19.5 nm, (c) 28.4 nm, (d) 30.4 nm.

Figure 2 : An active region at, from left to right, 17.1 nm (a) and 28.4nm (b).

Performing in accordance with the best expectations, EIT brings the paradoxal view of a very dynamical "quiet" Sun.

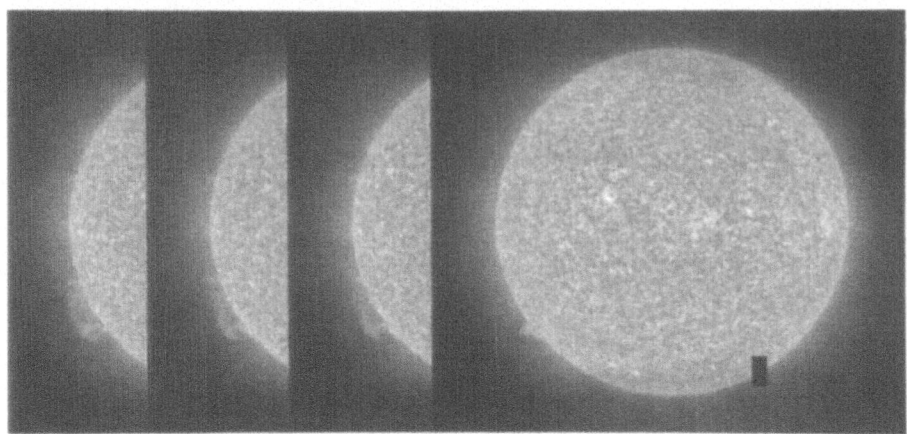

Figure 3 : The transition region at 30.4 nm, showing the emergence of an arch-prominence.

References :

[1]. Delaboudinière, J.-P. et al., 1989 : *EIT - Solar Corona Synoptic Observations with an Extreme-Ultra-Violet Imaging Telescope, The SOHO Mission*, ed. V. Domingo, ESA SP-1104, ESA, Noordwijk, 43

[2]. Clette, F. et al., 1995 : *EIT The Extreme Ultraviolet Imaging Telescope : Synoptic Observations of Small and Large-Scale Coronal Structures*, Proc. CESRA Workshop, Potsdam, "Coronal Magnetic Energy Releases", eds. A.O. Benz & A. Krüger, Lecture Notes in Physics, Springer Verlag, 251

[3]. Delaboudinière J.-P. et al., 1996 : EIT : Extreme-Ultraviolet Imaging Telescope for the SOHO Mission, *Solar Physics*, **162**, 291.

** EIT Science Consortium composition :*

J.-P. DELABOUDINIERE, G. E. ARTZNER, J. BRUNAUD,A. H.GABRIEL, J. F. HOCHEDEZ, F. MILLIER, and X. Y. SONG, *IAS, Orsay, France*;

B. AU, K. P. DERE, R. A. HOWARD, R. KREPIN, D. J. MICHELS and J. D. MOSES, *NRL, Washington DC,USA*;

J. M. DEFISE, C. JAMAR and P. ROCHUS, *CSL, Liège, Belgium;*

J. P. CHAUVINEAU and J. P.MARIOGE, IOTA, *Orsay, France*;

R. C. CATURA, J. R. LEMEN, L. SHING and R. A. STERN, *LPARL, Palo Alto CA, USA;*

J. B. GURMAN, NASA/GSFC, Greenbelt MD, USA, W. N. NEUPERT, *SEL/NOAA,Boulder, CO, USA;*

A. MAUCHERAT, F. PORTIER-FOZZANI, *LAS, Marseille, France;*

F. CLETTE,P. CUGNON, E. VAN DESSEL, J.-R. GABRYL, ORB, *Bruxelles, Belgium.*

OBJECTIVES OF A UKRAINIAN SOLAR-ORIENTED SATELLITE

I.P. KRYACHKO, L.N. KUROCHKA, V. TEL'NYUK-ADAMCHUK
Astronomical Observatory of Kyiv, Taras Shevchenko University
3 Observatorna street, UA-254053 Kyiv, UKRAINE

Abstract. The project is focused on the utilization of the Ukrainian space technology and industrial potential, within the framework of the National Space Agency of Ukraine activity. The initial stages of the project,are:
- concept formulation
- compilation of preliminary proposals,
- outlining of the set of appropriate scientific experiments to be made in space, matching both the demands of scientific experiments and the technical possibilities for their accomplishment, in light of the desirable orbit parameters, satellite orientation accuracy, and so on.

The project is to launch a multiaimed Sun-Oriented Automatic Universal Orbital platform designed and produced by the Ukrainian space industry and equipped by systems ensuring stable orientation of the satellite on the Sun. The platform includes an onboard control complex, electric power supply system, thermal control system, and a system for scientific data transmission to ground stations. The following are some of the mission parameters: total weigth 2000 kg; scientific payload 400 kg; active lifetime 2 y; altitude 500 km; inclination 57°; orientation accuracy 1 arc min; angular stabilization 0.005 /sec; recurrent period of the orbit, when the spacecraft is outside Earth shadow 20 days. The spacecraft will be launched by an Ukrainian rocket.

Scientific objectives include the study of different modes of solar oscillations; overall solar parameters variations, using observations in different wavelength bands. The project concept is based on an essential potential of Ukrainian solar physicists, and yhe achievement of the Ukrainian astronomer developed in the Crimea Astrophysical Observatory. Results have beenobtained in Kiev and Kharkiv Observatories, especially in the study of solar oscillations and specially the 160-minute oscillations.

Spacecraft preliminary scientific equipment includes the following:

1. DIFOS3 [1,2,3]. A Modern generation of a three-channel DIFOS panoramic photometers for recording 160-minute oscillations. DIFOS now is being modernized at the Main Astronomical Observatory of the Ukrainian Academy of Sciences. The first DIFOS version operated in Sun-like-a-star mode during the CORONAS-I space mission. The advanced version gives a panoramic image of the Sun. Spectral ranges are as follows: two channels are centered on the wavelengths 520 nm and 710 nm, and the third covers the whole spectral range from 400 to 1000 nm. DIFOS is capable of detecting the 1st to 3rd modes from the whole Sun and from its different parts.

Z. Mouradian and M. Stavinschi (eds.),
Theoretical and Observational Problems Related to Solar Eclipses, 233-235.
©1997 *Kluwer Academic Publishers.*

2. A seven-channel IR photometer [4] proposed by T. Kostiuk and D. Deming, wavelength longward of 1.6 mkm (to 200 mkm), can be used to great advantage for precise measurement of the solar photospheric gradient, which modulates total solar irradiance; to observe the solar flares in the IR emission range that is so effective when using the spaceborne instruments. For wavelengths below 35 mkm, instrumental sensitivity is adequate to detect the intensity signature of solar p-mode oscillations in a single 5 minutes of integration. For longer wavelengths, clear detection is expected using Fourier analysis of modest data sets.

3. SOLRAD [5]. The determination of solar diameter and its possible variability is one urgent task of solar physics. A problem arose again a decade ago while treating of long series of meridian observations, observations of the Mercury transitis across the solar disk, and so on. Groundbased solar diameter measurements made by a number of observers using different approaches estimate the solar radius in the interval from 961.2 arcsec to 959.2 arcsec. Such accuracy is insufficient now. Better accuracy can probably be achieved using astrolabe observations (French team) and solar eclipse observations near the totality limits (International Occultation Timing Assoc. teams). Several special ground and spacebased instruments have been proposed to measure the solar diameter with an accuracy of 10 mas or better. None of these has been launched in space. There have been several attempts to obtain such measurements by balloon. Spacebased experiment was also planned by the FSU in the Regatta-Astro project, but this project did not go through realized. Thus, we propose to include the instrument in the set of experiments to be fulfilled.

4. SUBCORONA [6,7] experiment is aimed at studying the subtelescopic features within the coronal macroscopic structures (coronal holes, streamers, and so forth etc.) and in coronal features (e.g. flares, prominences) as well as in spicules, surges, and so. Experiment is also important for understanding the problem of particle transport through the solar corona and for recording solar coronal transients. This goal can beachieved by recording and analyzing coronal images in both the Lyman and Thomson continuums. Basically the method consists in the using different electron density dependence laws in these two continuums. It must be emphasized that the corona radiation intensity in Lyman continuum is two orders of magnitude greater than that in the Balmer continuum. This investigation provides a way for improving the present standard model of the solar corona significantly and obtaining new information concerning solar corona features.

5. THE INHOMOGENEITY [8-13] experiment is to study macro- and micro-inhomogeneities of the structure, and its dynamics in both the active and quite solar regions on the solar disk as well as emission features such as flares, flocculi and so forth.

Macro-inhomogeneities of the structure and electron density may be studied successfully in the Lyman continuum because the spectral band background intensity in this range is much less than in the Balmer continuum; on the other hand, the emission intensity of the solar formations in the Lyman continuum is 130 times greater than that in Balmer continuum. Micro-inhomogeneities of the structure and electron density (along the line-of-sight) will be studied for the first time using a new method developed in Kiev University. These data are of importance for understanding origin and variation of the solar flare, as well as of other solar formations. Solar disk images will be obtained in the Lyman continuum, 89-91 nm, and no fewer than three images in the Lyman precontinuum, 91.2-93 nm, to study the origin and developement of flares, prominences, and coronal transients.

6.THE EMLY [14-16] experiment is aimed at studying the physical conditions in various solar emission formations. To solve this problem, the spectrum must be obtained between 90 and 122 nm. This band containt all hydrogen spectral lines and continuum of Lyman serie, as well as other emission lines. We are planning to use the lines profiles and their intensity for the analysis. These investigations will be useful in understanding the nature of solar flares, prominences, and so forth. Other proposals may be considered as well.

References

1. Gurtovenko E.A. et al. (1994) Photometer DIFOS for the study of solar brightness variations. *Solar Phys.*,**152**, 43-46
2. Lebedev N.I. et al. (1995) First results of the CORONAS-DIFOS experiment. Space observations of solar irradiance oscillations. *A&A*, **296**,.L25-L28
3. Kostyk R.I. et al. (1996) Helioseismological Coronas-Difos experiment. Space Science and Technology, .(in press).
4. Kostiuk T. and D. Deming. (1991) A solar infrared photometer for space flight application,. *Infrared Phys*, **32**, 225-233
5. Sofia S. et al. (1984) Solar Disk Sextant, *Appl. Opt.*, **23**, No 8,.1235-1237
6. Kurochka L.N., Matsuura O.T., Picazzio E. (1996) Subtelescopic inhomogeneities of electron concentration in the solar corona. *Solar Phys.* ,(in press).
7. Kurochka L.N., Tel'nyuk-Adamchyk V.V. (1996) Possibility to study subtelescopic structure in the solar corona. Theoretical and observational problems related to solar eclipses, *NATO workshop, Abstracts,1-5* June Sinaia, . 23
8. Kurochka L.N., Kiryukhina A.I. (1989) Micro- and Macroinhomogeneities in a quiescent prominence. *Hvar Obs. Bull.*,. **13**, No 1, ,(Proc. IAU Coll. 117. Dynamics of Prominence, Hvar.), .51-61
9. Kurochka L.N. (1991) Registration of macro- and microinhomogeneities of plasma density in astrophysical objects, *Kinematika i Fizika Nebesnych Tel*, , 7, No 4, .31-37.
10. Kurochka L.N., Kryachko I.P., Markova E. (1992) On the recording of regions with increased plasma density on the Sun, *Solar Phs.*, **139**, 275-277.
11. Kurochka L.M. . (1994) Method determination of electron density hydrogen plasma inhomogenities, Patent in Russian No 2019069, *Bull. izobretenij*, No **16**.
12. Kurochka L.M. (1995) The Balmer jumping natural hydrogen plasma, *Solar Phys.*, **156**, No 1, 65-71
13. Kurochka L.N., Matsuura O.T., Picazzio (1996) E. Determination of the electron density inhomogeneities of hydrogen plasma, *Astrophysics and Space Science*, (in press).
14. Kurochka L.N., Tel'njuk-Adamchuk V.V (1978) Calculation of line profiles of inhomogeneons solar formations, *Solar Phys.*, **59**, 11-19.
15. Kurochka L.N., Ribko L.B. (1978) The last observable line in hydrogen emission spectrum, *Solar Phys.*, **57**, 319-328.
16. Bruns A.V. et al. (1979) Spectroscopic investigations of solar active regions on Salyut-4, *Izv. Crimea Astrophys. Obs.*, . 59, 3-30.

A PROJECT FOR INFRARED OBSERVATIONS DURING THE TOTAL SOLAR ECLIPSES FROM THE CIRCUMTERRESTRIAL ORBIT

O. MARIS
Gravity and Space Sciences Institute
P.O.Box: MG-6, RO 76900 Bucharest - ROMANIA
e-mail: maris@roifa.ifa.ro

Abstract. On April 23, 1996, the PRIRODA international experiment was orbited. This experiment module is design ted to work together with the MIR orbital station. The Multichannel Infrared Spectrometer ISTOK-1 is one of the devices used in the experiment. ISTOK-1 measures the infrared Earth's emission spectrum in the 4-16 μm range, using 64 separated channels. The optical system and detector array are placed on a two-axis platform. The instrument is programmed, among other things, to orient platform to the Sun and investigate the medium infrared range. This paper presents the instrument characteristics and its possible utilisation for solar observations during total solar eclipses.

1. Introduction

The first observations of solar eclipses were performed in the visual spectrum. But the development of detectors in other spectral domains has enlarged the field of research. The infrared, which can provide much data on solar phenomena, is part of this development.

Ground-based infrared observations have some drawbacks:

- The Earth's atmosphere acts as a filter, absorbing certain radiation [1].
- Infrared radiation emitted by the terrestrial atmosphere overlaps that received from the Sun.
- Atmospheric dust (produced by significant natural phenomena such as major volcanic eruptions) distorts the measurements.
- Atmospheric water vapour and cloud-cover can weaken (and even blot uot complet) the signals.

All these drawbacks are avoided if the observations are made outside the atmosphere.

On April 23, 1996, a complex remote sensing module, called PRIRODA, was launched; and is now coupled to the MIR space station. The devices on PRIRODA were build by Russia, Germany, USA, Poland, the Czech Republic, Armenia, Ukraine, and Romania. The Romanian specialists from the Gravity and Space Sciences Institute participated in the construction of the Multichannel Infrared Spectrometer ISTOK-1. The main purpose of this spectrometer is to investigate the circumterrestrial medium. However, since the device has the possibility of beeing automatically oriented towards the Sun, it can perform measurements in the mean infrared (4 - 16 μm), whence arises the possibility of using it for certain measurements during the solar eclipses.

Z. Mouradian and M. Stavinschi (eds.),
Theoretical and Observational Problems Related to Solar Eclipses, 237-241.
©1997 *Kluwer Academic Publishers.*

2. Construction of the Spectrometer

Figure 1 plots the functional diagram of the ISTOK-1 spectrometer.

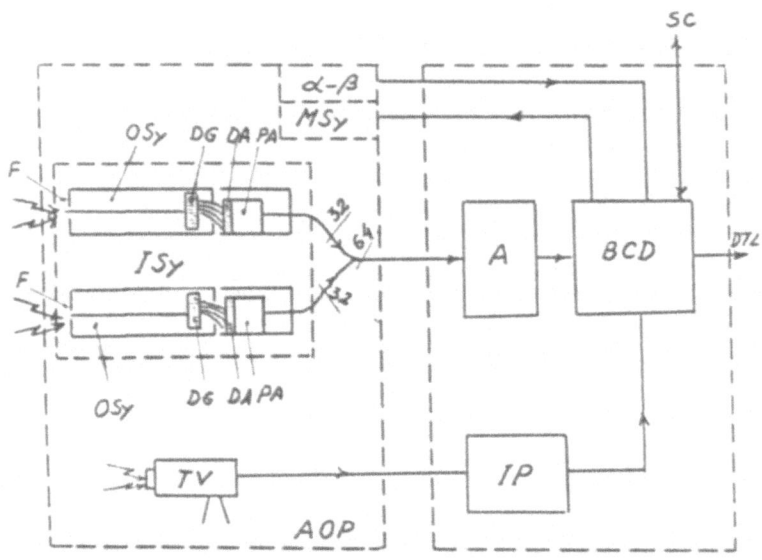

Figure 1. ISTOK-1 Functional Diagram

An automatic oriented platform (AOP), orientable around two perpendicular axes, is mounted outside the PRIRODA module. This platform carriesa telescopic system with infrared detectors (ISy) and a TV camera (TV) operating in the visual spectral range. Inside PRIRODA, the spectrometer ISTOK-1 also contains amplifiers for 64 spectrometric channels (A), a board computing device (BCD), and an image processor (IP) for the TV camera. The telescopic system on the mobile platform consists of two independent units: one for the spectral range 4 - 8 μm and other for the spectral range 8 - 16 μm [2]. Each unit consists of an optical system (OSy) placed in front of a diffraction grating (DG), which separates the spectrum into two ranges 4 - 8 μm, and 8 - 16 μm. The scattered radiation falls on a linear array of pyroelectrical detectors (DA) consisting of 32 detectors built on a LiTaO3 crystal. By command from Earth, it is possible to insert light cutting and light balancing filters (F) in front of the optical system. The signals from the 32 pyroelectric detectors are amplified by preamplifiers (PA) located directly behind the detectors. The amplified signals are taken over by BCD, which performs a preliminary processingon them. The spectometric information is then transmitted in blocks to the spacecraft, and finally to the Earth bases through a data transmission line (DTL).

The TV camera takes over the visual images provided by the same objective. In the "search and track" operating mode, BCD receives the Sun's coordinates from the spacecraft and the platform coordinates from the coordinate measuring system (α–β). This data is processed and the α and β displacement control is transmitted to the platforma motion systems (MSy). When the Sun reaches the TV camera field, the IP transmits the coordinates of the solar centre to BCD, which commands the platform motors until the telescopic system is pointed to the solar centre (with an accuracy of 80 arcseconds). The spectral measurements then are started and the platform (AOP) follows the Sun's motion on the celestial sphere.

The platform can be also commanded directly by the computer of the spacecraft, which transmits the orientation coordinates by means of BCD. In this way the platform can be oriented to any position, so any zone of interest can be scanned.

3. Device Performance

- Spectral range: 4 μm - 16 μm;
- Spectral resolution (average half-width of the spectral response function):
         0.15 μm - in the spectral range 4 - 8 μm;
         0.30 μm - in the spectral range 8 - 16 μm;
- Noise equivalent radiance: 10^{-5} W/cm$^2 \cdot$ sr $\cdot \mu$m;
- Interval between two spectral measurements: 1 s;
- Measurement accuracy (amplitude resolution): 0.5 %;
- Spatial resolution (spctrometer field of viewe): 6.5 x 26 arcmin;
- Spectral range of the TV-camera: 0.65 - 0.85 μm;
- TV-camera field of view: 20^0 x 15^0 and 2^0 x 1.5^0.

4. Possible Use

The calculations and the numerical values below correspond to the total solar eclipse of August 11, 1999.

In light of the specific features and characteristics described above, we feel that it might be used for several purposes:

a) *Scanning of the solar disk during partial eclipses, and spectral measurements in the range 4 - 16 μm on 64 distinct channels.*
Estimating the length of the spacecraft trajectory inside the penumbra (partial eclipse) [3] at about 9090 km, and the nominal velocity of the spacecraft at 7.8 km/s, there is a time window of 19 min 25 sec for the measurements.

b) *Measurements of the solar corona from the shadow of the total eclipse (band of totality), by scanning the circumsolar field.*
The maximum width of the totality band on the Earth's surface is 70 miles (112 km) [4], while the inclination of the spacecraft trajectory with respect to the line of centrality is 104 degrees. It follows that the maximum length of the spacecraft trajectory across the totality zone is 115 km. The maximum duration of the total solar eclipse for ISTOK-1 is 14.7 sec. Figure 2 plots the Sun's darkened contour during the

240

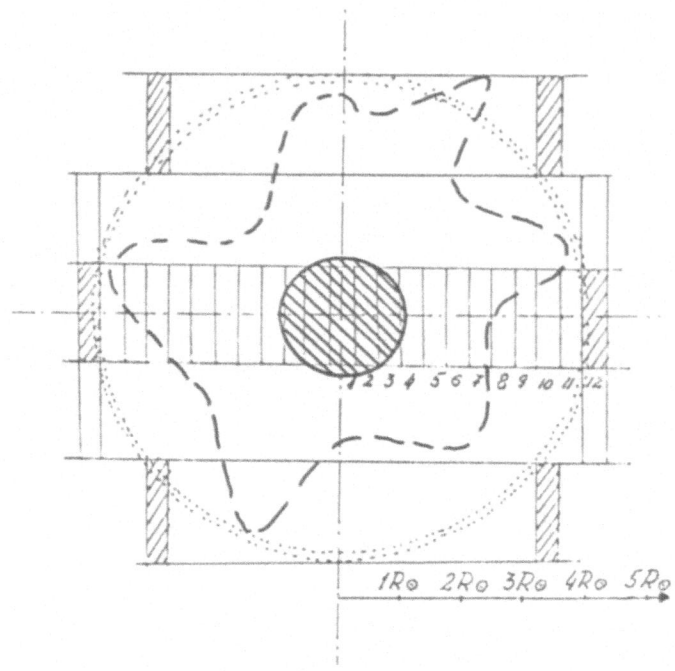

Figure 2. Measurement of the Solar Corona

eclipse, along with the corona, and the magnitude of the spectrometer pixel (1 pixel: 6.5 x 26 arcminutes). The measurements follow each other at an interval of 1. During the total eclipse, at then, most 14 measurements can be performed, so the whole corona cannot be scanned. It is also possible to cover the Sun's immediate neighbourhood by two exploring horizontal lines.

   c) *Observation of the hypothetical layers (rings) of circumsolar dust by measurements from the shadow (band of totality).*
On the basis of the previous estimates, and   examining   figure 2,it can be seen that, during the transit across the shadow, measurements can be made along a solar radius, from the disk edge up to a distance of 5 solar radii. Circumsolar dust layers have beendetected at distances of 3.5; 8.7 and 9.2 solar radii.

   d) *Measurement of the solar corona, in the visual domain, by means of a TV camera, during the eclipse (from the totality zone).*

   e) *Observations of the Sun from outside the eclipse, and measurements in the center of the solar disk, as well as automatic tracking of the solar disk center and measurements at 1 sec intervals.*

5. Conclusions

All these measurements can be performed in cooperation with the Russian coordinator of the spectrometer ISTOK-1, the *"P. N. Lebedev"* Institute for Physics of the Russian Academy.

We consider that in this way it will be possible to acquire data in the mean infrared 4-16 μm during the eclipse. Such data can be used for the analysis of the solar corona and of the circumsolar dust rings.

References

1. Deming, D., Jennings, D.E., Jefferies, J., and Lindsey, C. (1994) Physics of the infrared spectrum, in Rabin, D. M., Jefferies, J. T., and Lindsey, C. (eds.), *Infrared Solar Physics*, Kluwer Academic Publishers, Dordrecht, 933-963.
2. Dementyev, B. V., Ivanov, V. V., Kukin, S. G., and Maly, M. (1993) A new on-board multichannel infrared spectroradiometer for ecological monitoring and meteorological sounding of the atmosphere, in Belov, N. N., and Akopov, E. I. (eds.), *Optical Monitoring of the Environment 2107*, SPIE,. 74-80.
3. Espenak, F. (1987) Fifty Year Canon of Solar Eclipses: 1986-2035, *NASA Reference Publication 1178 Revised-1987*.
4. Espenak, F., and Anderson, J. (1996) Total Solar Eclipse of 1998 February 26, *NASA Reference Publication 1383, April 1996.*

EYE SAFETY DURING SOLAR ECLIPSES - MYTHS AND REALITIES

B. R. CHOU
School of Optometry, University of Waterloo
Waterloo, Ontario, Canada N2L 3G1

Abstract.:Safe eclipse viewing is a continuing problem for both astronomers and eyecare providers. Too often, the astronomer's enthusiasm for encouraging the public to observe an eclipse is countered by health care providers who consider that even the most minimal risk of eye damage is unacceptable. The challenge is to dispel misconceptions which are commonly held by both the public and many health authorities with regard to the eye hazards of observing solar eclipses.

1. Solar Retinopathy

Although the anterior structures of the intact human eye absorb ultraviolet radiation shorter than 380 nm, as well as infrared radiation longer than 1400 nm, the waveband between 380 and 1400 nm penetrates to the retina. [1] Consequently, unprotected or inadequately protected viewing of the sun can result in retinal burns. The damage arises from a photo-oxidative process referred to as photochemical retinopathy, which occurs in the retinal photoreceptors upon exposure to sunlight in the waveband 380 to 500 nm [2]. For unprotected viewing of the sun, the threshold exposure duration for clinically observable retinal damage is less than 60 seconds, although functional loss may arise from significantly lower exposures [2,3]. When an optical instrument such as a telescope is used, the threshold exposure time may be as brief as a fraction of a second[2]. Because of the discomfort glare due to the intense visible light from the solar disk, casual observers take brief frequently repeated glances at the sun during an eclipse. However, many subthreshold exposures during the course of an eclipse can accumulate to a threshold exposure for blue light damage. The high retinal irradiance level within the solar image also results in conversion of the incident light energy in the 380 to 500 nm waveband into heat, causing photocoagulation of the underlying retinal pigment epithelium when an optical instrument is used or if unaided viewing is prolonged [3]. This thermal damage also occurs with unprotected exposure to radiant energy at wavelengths between 500 and 1400 nm. Photocoagulation results in a permanently blinded area of the retina.

The risk of solar retinopathy is heightened because these injuries occur without pain sensation; the retina has no sensory innervation. In addition, the clinical signs and symptoms do not arise until between 12 and 48 hours after exposure [3]. Recovery of lost visual function is unpredictable and may be prolonged,

Z. Mouradian and M. Stavinschi (eds.),
Theoretical and Observational Problems Related to Solar Eclipses, 243-247.
©1997 *Kluwer Academic Publishers.*

depending on the relative levels of photochemical and thermal damage to the retina. Although many individuals recover their pre-eclipse visual acuity, tests of higher visual functions such as contrast sensitivity may show permanent losses; other individuals may be left with permanent visual losses ranging from a low degree of visual impairment to legal blindness [4]. This has an enormous potential socioeconomic impact, since most individuals affected by solar retinal injuries are children or young adults [4,5] whose post-injury vocational opportunities may be severely restricted.

2. Solar Filters

Although the safest method of observing the sun is by projection, many observers prefer to view the sun through protective filters. No international or national standard for solar filters exists, however it is recommended in Annex A of the European standard EN172: Personal eye-protection - Sunglare filters for industrial use that filters with scale number 12 to 16 are suitable for direct observation of the sun.[6] If this recommendation is followed, then a "safe" solar filter would have a maximum luminous transmittance of 0.0032% for the waveband 380 to 780 nm. The standard EN169: Personal eye-protection - Filters for welding and related techniques - Transmittance requirements and recommended utilisation [7] provides additional recommended specifications for filter transmittance in the ultraviolet (200 to 380 nm) and near-infrared (780 to 1400 nm) wavebands.

Table 1 lists a variety of filters which have proven safe for direct solar visual observation. Some are also suitable for photography. The spectral transmittance data were measured with a Cary 5 spectrophotometer with rear-beam attenuator accessory. Calculations for luminous transmittance (T), shade number (SN), effective-far-ultraviolet average transmittance (T-EFUV), near-ultraviolet average transmittance (T-NUV) and infrared average transmittance (T-IR)were carried out as specified in the ANSI Z87.1-1989 American National Standard Practice for Occupational and Educational Eye and Face Protection[8].

TABLE 1. Transmittance of selected solar filter materials

Material	T (%)	SN	T_{EFUV} (%)	T_{NUV} (%)	T_{IR} (%)
Questar glass	0.0024	11.8	4.44×10^{-5}	4.88×10^{-5}	0.394
1000 Oaks T1 glass	0.00084	12.8	3.46×10^{-5}	4.03×10^{-5}	0.160
1000 Oaks T2 glass	0.0016	12.2	2.76×10^{-5}	4.71×10^{-5}	0.036
1000 Oaks T3 glass	0.0053	11.0	2.76×10^{-5}	4.70×10^{-5}	0.075
1000 Oaks mylar	0.00025	14.1	4.27×10^{-5}	0.0011	0.0047
SolarSkreen visual grade	0.00013	14.7	5.47×10^{-5}	0.0034	0.0042
Welder's filter shade 12	0.0022	11.9	3.93×10^{-5}	3.52×10^{-5}	0.0049
Welder's filter shade 14	0.00023	14.2	3.42×10^{-5}	4.26×10^{-5}	0.0047
Floppy disk media	0.0023	11.8	4.10×10^{-5}	3.960×10^{-5}	3.79
CD-ROM	0.00024	14.1	3.42×10^{-5}	0.0001	0.0044
1000 Oaks Solar Shield2000	7.81×10^{-5}	15.3	3.14×10^{-5}	4.31×10^{-5}	0.117
Kodak Plus X (double)	0.0063	10.8	0.00016	0.0136	0.0112
Black color film (double)	7.39×10^{-5}	15.3	5.22×10^{-5}	4.05×10^{-5}	46.98

A safe filter can be made using a double thickness of silver-bearing black-and-white film that has been fully exposed to light and processed to maximum density. However, differences in processing methods and chemistry may result in considerable variation in final density of the emulsion. Thus a filter made of film may have a shade number varying between 11 and 16. Similarly, variations in manufacture of audio and data compact disks (CD or CD-ROM) may result in a wide range of density between disks. Compact disks with aluminum films that are semi-transparent in normal room lighting are unsuitable for use as solar filters. Floppy disk media have a marginally safe infrared transmittance and produce poor images of the solar disk.

Unsafe filters include any image-bearing photographic emulsion, chromogenic (non silver-bearing) black-and-white film, black processed color film, photographic neutral density filters, and polarizing filters. These materials transmit a high level of near-infrared radiation although the luminous transmittance may be very low. Transmittance data for a double thickness of black color film are presented in Table 1 as an example of this.

Although properly prepared smoked glass is an effective filter against both visible light, and infrared radiation, its use is strongly discouraged because of the difficulty in producing a uniform filter, and its extreme fragility.

3. Misconceptions about Eye Safety and Solar Eclipses

Every solar eclipse is presaged by announcements by the astronomical community, government agencies, and medical authorities about the dangers of unprotected viewing of the event. Although issued with the best of intentions, many of these announcements contain misinformation that can do more harm than good. This may arise because the writer lacks the necessary technical knowledge. Often the message is garbled by poor editing. In some instances, announcements may be worded negatively to discourage unsafe viewing practices. For example, in 1994, public service announcements by the American Academy of Ophthalmology forced the last minute cancellation of many school-based programs to observe the May annular solar eclipse because of the threat of legal action by panicking misinformed parents. There is often public confusion when advice provided by astronomers contradicts the advice from government or health care roviders.

There are several misconceptions about the safety of viewing a solar eclipse that recur frequently in the news media:

3.1 WHERE IS IT MOST DANGEROUS TO VIEW A SOLAR ECLIPSE?

News reports often leave the public with the impression that viewing an eclipse is dangerous only within the path of totality, and that looking at the sun from locations in the zone of partial eclipse or outside of the eclipse path is safe. Occasionally, announcements are worded to suggest the converse. Advisory messages should state clearly that unprotected or inadequately protected observation of the partly eclipsed sun is dangerous, but that it is safe to look at the total phase without a protective filter.

3.2 WHEN IS IT MOST DANGEROUS TO VIEW A SOLAR ECLIPSE?

Many public announcements state that looking at the eclipsed sun is always dangerous, even during totality. Some announcements have suggested that the danger to the eyes is higher because of solar radiation emitted during the eclipse that is not present at other times(!). While there is no physcial basis for this statement, it is a common misconception. Some warnings have also implied that the eye hazard exists even when the sky is overcast. During the eclipse of February 1979, a public service announcement to that effect resulted in children at many locations in Canada being kept indoors at school with windows covered, even though clouds and precipitation obscured the sun.

3.3 IS THERE A SAFE WAY TO VIEW AN ECLIPSE?

Advice on how to view a solar eclipse safely should include descriptions of both indirect projection methods and direct observation with protective filters. The common misconceptions are that there is no safe way to see an eclipse, and that the projection methods are the only foolproof safe methods. The use of filters made from a double layer of photographic film is sometimes mentioned, but without the warning that negatives bearing images and non-silver bearing emulsions are unsuitable for the purpose. Often, fear of incurring liability for eclipse-related eye injuries may cause announcement writers to discourage any attempt to observe the eclipse.

4. Public Education and Eye Safety

Information on eye safety and solar eclipses provided to the public is often flawed because of misinformation or conflicting messages from different authoritative sources. Co-operation with eyecare professionals to present a common message is important to the success of public education campaigns to prevent eclipse eye injuries. Experience has shown that messages that attempts to discourage eclipse viewing by "scare tactics" are generally ineffective, especially in the target population of adolescents and young adults [4,5]. Effective messages should be simple, attractively designed, and positive in tone, encouraging safe viewing practices. Any text should be at a reading comprehension level suitable for children of school age. The pamphlets and posters distributed in the Canadian provinces of Ontario and Quebec in 1994 are excellent examples.

To assist astronomers in preparing information on eye safety, Commission 46 of the International Astronomical Union set up a committee to provide resource information. Inquiries should be directed to the Chair, J.M. Pasachoff (e-mail: jmp@williams.edu). Transmittance data on safe solar filters can be obtained from the author (e-mail: bchou@sciborg.uwaterloo.ca).

References

1. Boettner E.A. and Wolter J.R. (1962) Transmission of the ocular media. *Invest Ophthalmol.* **1(6)**, 776-783.
2. Chou B.R. (1981) Protective filters for solar observation. *J. Royal Astron. Soc. Canada* **75**, 36-45.
3. Pitts D.G. (1993) Ocular effects of radiant energy. in D.G. Pitts and R.N. Kleinstein (eds.), *Environmental Vision: Interactions of the Eye, Vision, and the Environment*, Butterworth-Heinemann, Toronto, 151-220.
4. Chou B.R. and Krailo M.D. (1981) Eye injuries in Canada following the total solar eclipse of 26 February 1979. *Can. J. Optom.* **43**, 40-45.
5. Penner R. and McNair J.N. (1966) Eclipse blindness - Report of an epidemic in the military population of Hawaii. *Amer. J. Ophthalmol.* **61**, 1452-1457.
6. European Committee for Standardization (1994) EN172: *Personal eye-protection Sunglare filters for Industrial use.* CEN, Brussels.
7. European Committee for Standardization (1992) EN169:1992 *Personal eye- protection* - Filters for welding and related techniques - *Transmittance requirements and recommended utilisation.* CEN, Brussels.
8. American National Standards Institute (1989) ANSI Z87.1-1989 American National Standard Practic for Occupational and Educational Eye and Face Protection. ANSI, New York.

SOLAR ECLIPSES AS A VEHICLE FOR INTERNATIONAL ASTRONOMY EDUCATION

J. M. PASACHOFF
*Williams College--Hopkins Observatory, Williamstown,
MA 01267, USA*
jay.m.pasachoff@williams.edu

Abstract. The public's attention is drawn to astronomy whenever solar eclipse--partial, annular, or total--is visible, and we must take advantage of the opportunity to teach about the nature of science, the ability of astronomers to predict and analyze distant bodies and events, and the value of scientific research. We must also instruct people how to watch the partial and annular phases safely and that the total phase is not harmful.

In each eighteen-month interval, about one total solar eclipse and one annular solar eclipse occur. Further, eclipses that are no more than partial also occur, to a maximum possible five solar eclipses a year [1]. Though the bands of totality and annularity are only a few hundred kilometers wide at most, the area from which partial phases are seen are continent-wide. At the times when solar eclipses of any time are soon to be visible, the attention of the public is drawn to astronomy.

Though we professional astronomers know how exciting a total solar eclipse is compared with annular or partial eclipses, or the partial phases of total eclipses, the distinction is often lost to the general public. Yet we astronomers and educators have special opportunities to show our science and its value to an interested public at these times. Professors at local universities and colleges are often called by news media, and planetarium personnel are stretched to handle all the inquiries.

At the same time, however, the public often receives incorrect information. Often they are told that the eclipse is so hazardous that they will go blind if they look at it. Sometimes well-meaning professional groups of ophthalmologists spread such information, or wrongly state that the eclipse is best observed on television, and it has been my experience that such groups provide the information to the press so long in advance of the eclipse that it becomes widespread. Rarely is the point made clear to the public that there are advantages for them to observe the eclipse, such as the inspiration that it provides to individuals, especially to students. And rarely is it made clear in advance that the range of phenomena observed outdoors, including 360 degree changes in the sky, do not show well on television. Closeups, as available on television or in most photographs, do not stress the wide range of sky changes during totality (Figure 1).

Z. Mouradian and M. Stavinschi (eds.),
Theoretical and Observational Problems Related to Solar Eclipses. 249-255.
©1997 *Kluwer Academic Publishers.*

250

Figure 1. The corona in the sky during the 1995 total solar eclipse, observed from India. The dark shadow around the corona and the light, reddish sky near the horizon, make an awesome experience for the observer. (photo by Jay M. Pasachoff)

1. Observing Solar Eclipses

The public should be told months in advance of eclipses, by astronomers in each country from which the eclipse is visible, about the glory of solar eclipses and also about how to observe them safely. The distinction between the minutes of totality and the hours of partiality should be made. An evaluation of safe modes of observing eclipses should be made. Since most people hearing the message will see only partial phases--even those who live close to the band of totality--and since even those in the band of totality will see partial phases before and after the main event--methods of safely observing the partially eclipsed sun should be covered [2, 3]. Those methods include special solar filters (see the article by Ralph Chou, these proceedings), eyepiece projection from telescopes and binoculars, and pinhole cameras. Special solar filters can be purchased from various providers, or such filters can be made by fogging and exposing silver-bearing black-and-white film, and developing it fully. This latter idea is now beginning to suffer by the existence of black-and-white films without silver content, which provide insufficient safety. It should be stressed that even if the sun not visible through the

Figure 2. Pinhole crescents projected on the ground during the partial phases of a solar eclipse. (photo by Jay M. Pasachoff)

film, infrared heating can cause eye damage. But it is commonly thought that such damage can occur within seconds, which is far too low a period for observing the sun without optical aid. Instructions must be as clear as possible and an attempt should be made to anticipate misinterpretations. For example, it has often been cited that fogging and exposing x-ray film makes a good solar filter, and I recently discovered

252

that some people have been hauling out their old x-rays. Looking through where the bones are imaged, of course, provides no protection.

A favorite recommendation of mine for partial phases is pinhole projection. Usually, I describe how to make a pinhole camera out of a cardboard box or with pieces of cardboard. One must remember to stress that one looks down at the side of the box away from the end with the hole, or at the solid piece of cardboard on which the image is projected, rather than looking up at the sun through the hole! I like to point out that often one can take advantage of the natural pinhole camera formed by the interstices among the leaves of a tree (Figure 2).

Figure 3. The partial phase of an eclipse observed through a suitable solar filter. (photo by Jay M. Pasachoff)

2. The Importance of Correct Information

All too often, I have found misinformation given out to the general public, and then incorrect responses as a result. The most tragic circumstance occurs when students are prevented by teachers or school boards from going outside during eclipses to see the phenomenon. I have run into such situation several times, notably in Brandon, Manitoba, Canada, during the 1979 total eclipse and in Williamsburg, Virginia, USA, during the 1984 annular eclipse. In spite of the pleas of the science teachers, who were knowledgeable, higher ups prevented students from viewing the eclipse. In the Canadian case, students were locked in basement rooms on the side of the school building opposite that of the sun.

What happened later, I ask, when the eclipse was over and the students went home. What if they met friends or relations who said, "Did you see the eclipse? Wasn't it glorious." To their response that they were told in school that they couldn't see it because it was hazardous, and that they might have gone blind, they would be told that that information was ridiculous and that the eclipse was wonderful.

The real problem then comes later on, when the teachers or other adults tell these students, "Don't smoke. Don't take drugs." And, newly, "Don't have unprotected sex, which can lead to AIDS." The students will have lost faith in the accuracy of the messages they have been given. So the effect of receiving false or misleading information about the hazards of eclipses can be major and long-lasting.

3. International Efforts on Eclipses Education

As Chair of the Working Group on Eclipses of the International Astronomical Union, I have tried to distribute sample materials in different languages to people in different countries, to aid them in obtaining correct publicity about forthcoming eclipses. Members of the Working Group include Iraida Kim of Moscow, Yoshinori Suematsu of Japan, Jagdev Singh of India, Fred Espenak of the USA, and V. Rusin of Slovakia, assisted by meteorological consultant Jay Anderson of Canada.

A subcommittee of Commission 46 on the Teaching of Astronomy of the International Astronomical Union has recently been formed to take advantage of the public education opportunities of eclipses and to serve the public by providing information. I chair the subcommittee, which includes Julieta Fierro of Mexico and Ralph Chou of Canada. For the 1995 eclipse, for example, we were able to supply information for various countries from Canada to Vietnam to Guam, where a partial eclipse was visible. For the 1998 eclipse, we are already providing information in Spanish and in Portuguese to colleagues in South American countries, to aid them in providing materials.

4. Some Eclipses Observed

In my experience, there is no direct correlation between the technical level of a civilization and the amount of correct eclipse information given out. At the 1973 eclipse at Lake Loiengalani, Kenya, for example, though we were with primitive peoples, they were receptive to our lessons. For example, we arranged to sound the mission church bells during totality, so that people would know when it was safe to watch the eclipse with unaided eye.

At the 1974 eclipse in Australia and the 1979 eclipse in Canada, though, no amount of outreach we tried could convince people that schoolchildren should be encouraged to see the eclipse rather than prevented from seeing it. In Manitoba, Canada, the local astronomy professors had tried to reason with the local authorities, without success, and hoped that my appearance as an outside expert could tip the scales. But we did not succeed, and students during totality remained locked away from the eclipse. From our vantage point on a balcony at Brandon University, we could see, as totality began, cars turning on their headlights and continuing to drive.

At the 1984 annular eclipse in Williamsburg, Virginia, USA, no amount of explanation from the high-school teachers or from me would persuade the school board, a few days before the eclipse, to allow the students to watch it, even with the teachers' close guidance and supervision. In the event, it was rainy in Williamsburg. I wound up observing this eclipse from Picayune, Mississippi, where those of us in the town green were able to guide those gathered there.

At the 1994 annular eclipse that crossed the United States, things seemed a bit more hopeful. I observed the eclipse from the centerline in New Hampshire (Figure 3), with a satellite television truck broadcasting reports back to their Boston station. So I was able to provide information on how to watch the eclipse safely. With the help of the Smithsonian Astrophysical Observatory, I prepared a videotape that was distributed to television stations and others on how to make a pinhole camera; this tape remains available for future eclipses.

The situation in India was noticeably different in 1995 from that in 1980, when people seemed more scared of the eclipse and went indoors during its entire duration. In 1995, efforts by planetarium personnel and other scientists and educators had convinced substantial numbers of people that the eclipse per se was not harmful.

5. Conclusion

We have a substantial job to do in public education in many countries around the world for the next few eclipses, and we should be organized and start early. The public is receptive and interested, and we must prepare and distribute accurate information on not only the hazards but also the advantages of observing the eclipse, covering not only professional research but the personal interests of the general public.

The Working Group on Eclipses of the International Union's solar commissions and the subcommittee on solar eclipses of the IAU's commission on the Teaching of Astronomy stand ready to provide materials and otherwise help with public education.

My work at total eclipses has been supported by the National Science Foundation's Atmospheric Sciences Division, Astronomy Division, and Education Division, most recently through grants ATM-9005194, AST-9014889, USE-9050643, ATM-9207110, DUE-9351279, AST-9512216, and by grants from the Committee on Research and Exploration of the National Geographic Society. Our expeditions from Williams College, and student participation, have also benefitted by grants from the W. M. Keck Foundation and from the Brandi Fund and the Safford Fund.

References

1. Golub, L., and. Pasachoff, J. M. (1997). The Solar Corona (Cambridge University Press).
2. Pasachoff, J. M., and Covington, M (1993). The Cambridge Eclipse Photography Guide (Cambridge University Press).
3. Pasachoff, J. M., and. Menzel. D. H (1992) A Field Guide to the Stars and Planets, 3rd ed. (Houghton Mifflin Co.).

VARIATIONS OF GEOPHYSICAL AND BIOLOGICAL PARAMETERS OBSERVED AT A SOLAR ECLIPSE

I. PREDEANU
Astronomical Institute of the Romanian Academy
Str. Cutitul de Argint 5, 75212 Bucharest 28, Romania

1. Introduction

A total solar eclipse is accompanied by specific changes in the terrestrial magnetosphere, ionosphere and atmosphere. Certain effects also appear in several terrestrial parameters at ground level: air temperature, geomagnetic field and radio wave propagation all exhibit clear variations at the occurrence of the eclipse. At the same time, during a total solar eclipse, variations have been noticed in certain vital activities and physiological parameters of different organisms (insects, marine animals, sheep, humans).

2. Geophysical Effects

As the Moon's shadow passes through the Earth's atmosphere, it produces cooling and induces gravity waves. The temperature, barometric pressure, wind speed and direction, dew point, and humidityall change during the totality. Therefore, the following have been observed:
    - a significant reversal of the radiation balance, and variations in electrical activity in the lower atmosphere [1],
    - a sudden increase in wind velocities during totality [2],
    - a nonlinear drop of $3°C$ in atmospheric temperature[2],
    - a very significant drop of $20°C$ in soil-surface temperature, whereas the drop in the earth's immediate boundary layer is about $2°C$ [1].

The shadow of the Moon, traveling at supersonic speed, produces a thermal shock wave inducing pressure oscillations with periods from 2 min to 90 min and amplitudes between $0.12hPa$ and $20hPa$. The gravitational disturbances observed during the total solar eclipse of July 11, 1991, have been explained by the atmospheric pressure variations [3]. Wavelike structures in ionospheric parameters (the total electron content, the critical frequencies of the layers F2 and E, and so forth) have been also observed; the periods were from 20 min to 3 hours, depending on the distance from the central line of totality zone [4].

The solar occultation reduces the critical frequencies of the E and F2 layers, a decrease of the total electron content of F2 layer, a thickening of the F2 layer, and a rise in the F2 layer itself [5], [6], [7]. The decrease of the X-radiation related to a total solar eclipse was evaluated at about 10-20 per cent [8].

Z. Mouradian and M. Stavinschi (eds.),
Theoretical and Observational Problems Related to Solar Eclipses, 257-260.
©1997 Kluwer Academic Publishers.

During the eclipse, the amplitude of the signal at HF (10 MHz) and the field strength at LF (164 kHz) were observed to increase. The D region concentration gradually decreases and tends to disappear as the Sun is gradually coverd. As the D region ionization decreases, the VLF reflection height and the observed VLF phase increase [9]. Changes have also been in the phase height and absorption at 2.2 MHz [10], in atmospheric and magnetospheric radio noise. Radio wave propagation in the short, medium, and long wave bands is being improved considerably, especially in the short wave band [11].

Besides these, before the start of the eclipse, some absorption events occurred in the short and medium wave bands. These were explained by the ionization of the radiation reflected from the surface of the approaching Moon [11].

Also, enhanced waves have been noticed in geomagnetic data a few hours before the optical eclipse. These oscillations were assumed to be due to a corpuscular eclipse which precedes the optical effect [4].

The effect of eclipses on geomagnetism is believed to be due to the shielding of the solar X and UV radiation by the Moon. This decreases the ionospheric electron density and causesan additional current system, Sq. This additional Sq is in the opposite direction to the normal Sq, and depends on the magnitude of the eclipse, the altitude of the Sun, the position of the observer, the totality duration,and the level of geomagnetic activity. The eclipse variation curve can be obtained by subtracting the disturbed field curve for the non-eclipse zone from that for the eclipse zone [5].

Geomagnetic effects were reported during the annular eclipse of April 19, 1958 [5]; the solar eclipse of February 16, 1980, in India [12]; the solar eclipse of July 31, 1981, in the Bering Island, when the geomagnetic field full vector modulus decreased by about 10nT [13],and the total solar eclipse of June 11, 1983, in Papua when the horizontal component H decreased by 3.6nT, the declination shifted estward 0.6', and the vertical component Z increased by 5.0nT [14].

3. Biological Effects

During a total solar eclipse, changeshave been observed changes in animal behaviour, in terms of variations incertain vital activities and physiological parameters of different organisms.

Anomalous behaviour of adults, small babies and large dogs was observed at the total solar eclipse of 31 July 1981, in Kichera (Eastern Siberia) [15]: when the last quarter of the Sun started to be eclipsed, the temperature dropped, people became nervous and the babies began to cry, even if they were at their mothers' hands and could feel safe. When the Sun was a sliver, just before the second contact, dogs began to bark in fear. Some of them crawled on the ground to the tourists' tents and howled at the Sun and Moon. An interesting thing is that: Just at the same time, the fish in a large lake in southern Bohemia jumped out of the water with aloud splattes. This did not repeat itself later.

Cultures of bacteria suspended in saline solution,exposed for 30 and 60 minutes to solar radiation on the day of the solar eclipse, and subsequently inoculated in Nutrient broth and Mueller Hinton broth were a little turbid. At the same time, the control cultures inoculated by bacterial suspensions maintained in darkness for similar duration showed normal growth characteristics [16].

Marine zooplanktonic organisms migrated to the surface during the period of the eclipse [17]. Light and gravity are considered to be the two most important factors controlling this vertical migration.

The light-oriented activity of adult, pupae, and larvae, of diurnal and nocturnal mosquitoes, are reversed during a total solar eclipse [18].

The metabolism of the fish Anabas Testudineus drop significantly, on the day of a solar eclipse, and the respiratory rate was low, in spite of the fact that there was a rise in temperature [19].

Certain physiological parameters of pregnant sheep have been modified as follows [16]:

- The pulse and respiratory rates of sheep exposed to sun decreased during the eclipse,along with the hemoglobin concentration;

- The pulse rateof sheep exposed to sun until the start of the eclipse increased while the respiratory rate decreased;

- There were no changes in pulse, respiratory rate, or hemoglobin concentration, in sheep kept in the shade all the day.

The effects of a total solar eclipse on the functioning of the cardiovascular system were noticed in 13 patients (12 males and one female). Fourteen normal healthy individuals, all men, served as control group [20]. Changes during an eclipse are noticed only in heart patients: six have suffered changes in pulse rate, 12 in blood pressure, 8 in rate of sedimentation of erythrocytes (ESR), and 2 in E.C.G.

Observations of certain physiological parameters were made on a group of nine young students during a total solar eclipse[21]. Statistically significant alterations were found: fall in the leucocyte count, with a rise in the total platelet count and a reduction in the coagulation time.

A total solar eclipse is a peculiar case of the New Moon. The human being seems to be influenced by the variations of external factors related to Sun-Moon-Earth configurations. Some researchs [22] has revealed that, the temperature, atmospheric pressure, and geomagnetic activity are lower more frequently at New Moon. After the New Moon, the rainfall, temperature and sudden cardiovascular mortality increase. Weather frontal disturbances and hospitalizations for cardiovascular diseases also increase after the New Moon.

The total solar eclipse in 1999 will be an opportunity for making special observations of several physical and biological parameters, and for seeking their possible relationships.

References.

1. Kulshrestha, S.M., Chatterjee, K. and Srivastava, G.P. (1981) Meteorological observations in connection with the total solar eclipse of 16 February 1980, in Ref.23, 91-96.
2. Lalrosanga, R. (1981) Observation of total solar eclipse on 16 February 1980, in Ref.23, 108-109.
3. Ducarme, B., Sun, H.-P., d'Oreye, N., Van Ruymbeke, M. and Mena Jara, J. Interpretation of the tidal residuals during the July 11, 1991 total solar eclipse. (in press).

260

4. Jakowski, N. and Bettac, H.-D. (1983) The ionospheric response to the solar eclipse of 26 February 1979 observed in Havana / Cuba, *Phys. Solariterr.*, Potsdam, **20**, 110-116.

5. Solar Eclipse Group For Coordinated Observations (1977) A summary report on the coordinated optical, radio, geomagnetic and ionospheric observations of the solar eclipse of September 22, 1968 in Sinkiang, *Chinese Astronomy*, Pergamon Press, **1**, 105-129.

6. Chandra, H., Sethia, G., Vyas, G.D., Deshpande, M.R. and Vats, H.O. (1981) Ionospheric effects of the total solar eclipse of 16 February 1980, in *Ref.23*, 57-60.

7. Indyukov, A.E., Taschilin, A.V., Fligel', M.D. (1986) Effects of the solar eclipse on the outside ionosphere (in Russian), *Geomagnetizm i Aehronomiya*, Tom 26, **4**, 557-562.

8. Ivanov-Kholodnyj, G.S. and Kishcha, P.V. (1987) The source of the ionization of the ionospheric E-layer from observations during solar eclipses, (in Russian), *Geomagnetizm i Aehronomiya*, Tom 27, **2**, 307-309.

9. Dixit, P.S., Rao, P.K., Bhonsle, R.V., Sethia, G., Deshpande, M.R. and Chandra, H. (1981) Phase and field measurements in VLF, LF and HF regions during the solar eclipse of 16 February 1980 - Preliminary results, in *Ref.23*, 67-70.

10. Purkait, N.N. and Das Gupta, M.K. (1981) Effects of the solar eclipse of 16 February 1980 on the change in phase height and absorption at 2.2 MHz, in *Ref.23*, 74-76.

11. Sen, A.K., Saha B., Trehan S.K., Sekhar De, S., Saha, S.K., Datta, R.N., Chatterjee, S.K., Sehra, J.S., Das Gupta, M.K. and Sen, S.K. (1981) Radio wave propagation during the total solar eclipse of 16 February 1980, in *Ref.23*, 77-79.

12. Rangarajan, G.K. and Murty, A.V.S. (1981) Solar eclipse effect on short period geomagnetic field variations, in *Ref.23*, 51-52.

13. Kolokolov, L.E. and Russkikh, B.S. (1983) A decrease of the geomagnetic field strength modulus during the solar eclipse of 31 July 1981 (in Russian), *Issledovaniya po geomagnetizmu, aehronomii i fizike solntza* M.:Nauka, **66**, 108-110.

14. Liu Chang-fa (1986) The effect on the geomagnetic field of the total solar eclipse in Papua New-Guinea, *Acta geophys. Sin.*, Tom 29, **6**, 631-635.

15. Kotrc, P. (1996) Private communication.

16. Uppal, P.K., Kamal Kishore, Balwar, L.M., Prasad, V.S.S, Banerjee, R. and Ravindra Kumar (1981) Effect of solar eclipse radiations on sheep and microbial growth, in *Ref.23*, 131.

17. Bhattacharyya, S.S. (1981) Effect of solar eclipse on marine animals, in *Ref.23*, 120-121.

18. Anasuya, D.R. and Gopinath, K. (1981) Reversal of light oriented activity in mosquitoes with different circadian rhythms during a total solar eclipse, in *Ref.23*, 115-119.

19. Datta Munshi, J.S., Ghosh, T.K. and Biswas, N. (1981) Effect *Ref.23* of solare eclipse on the circadian rhythm of bimodal oxygen uptake in a climbing perch "Anabas testudineus" (Block), in *Ref.23*, 122-123.

20. Rada Krishna Murthy, P., Chakrapani, T. and Saraswathi, P. (1981) Effect of total solar eclipse - 1980 on the functioning of cardiovascular system, in *Ref.23*, 124-125.

21. Shanmukhappa, N.J. and Narayana Setty, C.S. (1981) Observation on certain physiological parameters during total solar eclipse of 16 February 1980 ad Davengere, in *Ref.23*, 126-130.

22. Sitar, J. (1993) On the semilunar periodicity in meteorologic and geomagnetic indices and in cardiovascular disease, Poster , September 1993, Košice, Slovac Republic.

23. S. K. Trehan, M. Dhara and J. Saketharaman (eds.) (1981) *Observations of Total Solar Eclipse of 16 February 1980*, Indian National Science Academy, New Delhi, 134 .

LIST OF POSTER PAPERS [1]

Session 1

Considerations on the Participation of Cluj Astronomical Observatory in the Solar Total Eclipse of February 15, 1961.
    G. D. CHIS, E. RADU, T. OPROIU
The Corona Electropolarimetry During November 3, 1994 Solar Eclipse.
    V. I. KULIJANISHVILI, V. KAKHIANI
Observation of the 1961 Total Solar Eclipse in Bulgaria.
    M. MADJARSKA, V. N. DERMENDJIEV
Structure of the White-Light Corona on the Eclipse of October 24, 1995.
    E. MARKOVA, M. BELIK, T. SYKORA, M. SOLC, J. SOLC, L. SMELCER
Results of the 1994 and 1995 Eclipse Observations.
    J. SYKORA, P. AMBROZ, P. KOTRC, M. MINAROVJECH, T. PINTER,
    J.RYBAK, M. RYBANSKY
The White-Light and Emission Coronae of the October 24, 1995 Eclipse.
    V. RUSIN, L. KLOCOK, M. MINAROVJECH, M. RYBANSKY
Possible Changes in the Solar Corona During the November 3, 1994 Eclipse.
    V. RUSIN, E. MARKOVA, M. BELIK
Observational results of the 1961 eclipse in Romania.
    E. TIFREA
Belgrade Observatory Experiences in Total Solar Eclipse Observations.
    I. VINCE, A.KUBICELA, J. ARSENIJEVIC

Session 3

Observations of the Emissions of the Upper Atmosphere During the Solar Eclipses.
    Y. V. PLATOV
Dark Saddle Structures in the X-Ray Corona.
    B. P. FILIPOV
Prominence Ejection into the Corona on September 24,1991.
    L. ALTAS, A. DUZGELEN
Numerical Simulation of a Prominence Formation in a Eeutral Sheet.
    C. DUMITRACHE, D. M. SURAN
The Length of Filaments as Function of Their Inclination on the Solar Parallel.
    C. DUMITRACHE
Observations and Analysis of Surges in Chromosphere and Lower Corona.
    P. KOTRC, B. SCHMIEDER, M. KARLICKY, P. HEINZEL

Session 4

Some Non-Traditional Ways of Solar Eclipse Observations.
R. A. GULYAEV
The Study of the Fast Changes in the Coronal Structures.
M. LORENC, M. MINAROVJECH, M. RYBANSKY
An Interesting Experience on the Emission Corona Observation Through a Narrow-Band Filter at the Thin Clouds Weather Conditions.
J. SYKORA
About a Possibility of Astrophysical and Geophysical Researches on a Board of Ballistic Rocket.
Y. V. PLARTOV, S. A. PULINETS, I. I. VELICHKO

Session 5

Intended Experiment of the Group from Observatory Upice for the 1999 Eclipse.
M. BELIK, E. MARKOVA
Preliminary Programme for the 1999 Total Solar Eclipse Observation in Bulgaria.
V. N. DERMENDJIEV, D. MISHEV, M. MADJARSKA
Statistical Analysis of Meteorogical Data Along the Path of the 1999 Total Solar Eclipse in Bulgaria.
I. T. ILIEV, V. I. TSANEV, D. MISHEV, V. N. DERMENDJIEV
Appropiate Observing Sites and Local Facilities for the 1999 Solar Eclipse in Turkey.
A. OKTEN, L. ALTAS
Reobservation of Subtelescopic Coronal Condensation in the Solar Eclipse of February 26, 1998.
E. PICAZZIO, O. T. MATSUURA, L. M. KUROCHKA, F. I. FOGLIANO
Observation of Polarization of Corona and Flash spectrum at the Total Solar Eclipse of August 11, 1999.
L. C. POPOVIC, I. VINCE
Experiments (Planned) for the 1999 Eclipse.
V. RUSIN, M. RYBANSKY
Belgrade Observational Program for the Total Solar Eclipse of August 11, 1999.
I. VINCE, A. KUBICELA, J. ARSENIJEVIC, L. POPOVICI
Solar Corona and Wind Built-up by Spicular Ejections.
O.T. MATSUURA, E. PICAZZIO
The Solar Eclipse on the 8th/21th August 1914 Reflected in Romanian Media of that Time.
E. BOTEZ

---

[1] The Poster papers are published in *ROMANIAN ASTRONOMICAL JOURNAL* 1996 volume 6 Sup.

NOTES ON GENERAL DISCUSSION [1]

Coronal heating: We need more information about the location of the structures in which heating occurs. For example, does it occur in current sheets or sheaths surrounding flux tubes? What are the size and temperatures of these structures. (P. Ulmschneider)

Polarization measurements: We need more imaging of polarized light from coronal features such as streamers, coronal holes, and current sheets using white light. Measurements in white light provide a relatively pure measurement of electron density without the use of any arbitrary parameters. (J. Sykora)

We need to reconsider the question of whether the prominence mass is really much greater than the mass deficient in the surrounding cavity as claimed by Tandberg-Hanssen and Sato. New techniques are now available which can make much better measurements in this regard. (S. Koutchmy)

Is there a connection between streamer densities and coronal heating? Could there be a condensation in reverse? (S. Koutchmy)

Could spicules supply the mass of a prominence as proposed by Athay and Klimchuk? (P. Ulmschneider)

Careful observations of the polarization of prominence cavities and the overlaying helmet streamers could help to determine the physical relation of the streamer and cavity to the prominence. (E. Hiei)

Time variations in the cavity density suggest that there is no close relation between the cavity density and the prominence density. (Z. Mouradian)

There is a problem with past eclipse observations of streamer and cavity structures which have used radial gradient filters. These filters reduce the photon influx and thereby throw away much of the data that could be useful. (J. Sykora)

The multiple exposure technique has less of a problem in this regard, and therefore is, in my opinion, superior to the use of radial gradient filters. (E. Hiei)

CCDs are another possible way to capture all the information contained in the photon flux, at least in principle. (S. Koutchmy)

There is a problem with CCDs because they have a very nonlinear response which leads to distortions and makes precise quantitative work difficult. (P. Fox)

CCDs allow detection of waves, which is not so easy to do with photographic techniques. (S. Koutchmy)

We need to address the problems of obtaining good photometry and good standardization with multiple sites . (Z. Mouradian)

Three-dimensional structures complicate the problem of detecting waves. Perhaps polarization observations could be used to separate out three-dimensional structures and make it easier to see waves (I. Kim)

Regarding heating in spicules, should we look for heating in their outer layers or in their cores? (E. Hiei)

Probably we should look in both regions. However, most theories predict heating only in the outer layers and have considerable problems explaining how heat can be transferred to the interior. The same problem also occurs at streamer boundaries. (P. Ulmschneider)

Temperatures inferred from the red and green coronal lines might be able to determine the location of heating. (E. Hiei)

There is a paper by P. Foukal (Solar Phys. 43 327) which reports evidence for heating at the surface of coronal archs of active regions. (Z. Mouradian)

Providing an archive of previous publications and observations would be very helpful (P. Fox).

I have started an archiving project, but it is not yet finished. (J. Pasachoff)

An archive would be useful for funding, and it would be especially useful if it could be accessed from the Internet. (S. Koutchmy)

It should not be forgotten that there are already extensive reviews of eclipse observations available in the literature (Z. Mouradian).

We need to discuss the impact of eclipse science on SOHO, as there is a good potential for special SOHO campaigns with LASCO and UCVS to augment ground based eclipse observations. (F. Clette)

Previous rocket and satellite programs do not make previous ground base observations obsolete. On the contrary, eclipse observations should become more valuable because they will enhance the rocket and satellite programs. (P. Fox)

An example of such a collaboration is the use of the Newkirk K-coronagraph at Mauna Loa, during the 1991 eclipse to calibrate spacecraft instruments and also to calibrate the non-eclipse polarimetric coronagraph. (J. Pasachoff)

The Yohkoh team already has a special program for collaboration with ground based eclipse observations. (E. Hiei)

ESA is willing to help us if there are some concrete programs or proposals for observations that we can suggest. (M. Stavinschi)

SUMMARY STATEMENTS OF SESSION CHAIRS [1]

Koutchmy (session "Small and large scale models of coronal structures"): There are two recommendations I can make for the future. One is that we should use the latest advances in CCD technology to obtain high resolution observations of the density structures in the vicinity of prominences. Improved observations should increase our understanding of the physics of the prominences and the cavity which surrounds them. The second recommendation is that we collaborate with the SOHO PI's to select one or more regions on the Sun prior to the eclipse as a priority target.

Maris (session "Small and large scale models of coronal structures"): Polarization observations continue to reveal new information about the corona and are proving very useful in understanding the structure of polar plumes. Collaborations with spacecraft including not only SOHO, but also Yohkoh and Ulysses, should be a priority for the 1999 eclipse.

Ulmschneider (session "Low temperature structures in the coronal environment"): Eclipse observations of prominences may help us decide between the various prominence models that were discussed at this meeting. Observations of chromospheric fibrils and magnetic field measurements using the Hanle effect have already eliminated the simple Kippenhahn and Schluter model, that was the standard model 20 years ago. Observations of structures in the coronal cavity might serve to eliminate several of the new models that have been proposed in recent years by increasing observational constraints on the topology of the magnetic field structure in and around the prominence.

Hiei (session "Specific problems of solar eclipse observations"): Emission-line observations of the corona have the potential to test some of the mechanisms proposed for heating the corona. However, further improvements are needed in the accuracy of the measurements which are made. Collaboration between theorists and observers help to increase the scientific value of eclipse observations.

Pasachoff (session "Tasks for total solar eclipse of 11 August 1999"): We must get weather assessments along the path. Preliminary results show increasing chances of clear weather as the eclipse progresses to Hungary and Bulgaria and even better chances in Turkey and beyond. We must assess whether one can make satisfactory observations from the Bucharest Observatory itself or whether the advantages of being in the mountains outweigh the logistic difficulties. As Chair of the IAU Working Group on Eclipses, I will be writing to solicit National Representatives in the various countries crossed by the path, with a view to getting official help in such things as waiving customs duties.

Forbes (session "Public education at eclipse and eye safety"): Although total eclipses provide a special opportunity to excite public interest in solar astronomy, there has been a trend in some countries to ignore eclipses. Worse still, some educational institutions" but would rather say "some organizations grossly misrepresent the safety issues involved; for example, optometrists' organizations sometimes overstate the risks without evaluating any benefits in observing eclipses. In addition to carrying out our scientific goals, I would hope that all of us could help promote public awareness of solar eclipses.

Sykora ("Poster Session"): The poster topics are so heterogeneous that it is difficult to summarize all of them. However, I can state that I was impressed with the results that continue to come out of past eclipse campaigns such as those in 1961 and 1994. This suggests to me that we still need to fund the analysis of previous eclipse observations as well as plan for new campaigns. There were very few posters which proposed specific projects for the 1999 eclipse, so we still need to develop precise plans that will take advantage of recent developments in technology.

---

[1] Notes by T. Forbs

AUTHOR INDEX

SUBJECT INDEX

The manufacturer's authorised representative in the EU is Springer
Nature Customer Service Centre GmbH, Europaplatz 3, 69115 Heidelberg,
Germany. If you have any concerns regarding our products, please
contact ProductSafety@springernature.com

Printed and bound by CPI Group (UK) Ltd, Croydon, CR0 4YY
29/04/2026
02099472-0003